Particle Mechanics
The Theory of Energy States

Welcome to
21st Century Physics

Dannel Roberts

Lions & Tigers & Bears, Publishing Inc.® / USA

Copyright © 2004 by Dannel Roberts
All rights reserved

No part of this publication may be produced in whole or in part, or stored in a retrieval system, or transmitted in any form or by any means, electronic, mechanical, photocopying, recording, or otherwise, without written permission of the publisher.

Cover Art and Book Design by Linda S. Stolte

Illustrations by Henry Thomas and Dewaine Thomas

ISBN 1-893459-05-5

Published by Lions & Tigers & Bears, Publishing Inc.
Printed in the United States of America

MMIV

Contents

Preface   v

Law 1:   Energy Can Move Through Matter   3

Law 2:   Energy Can Be Reflected By Matter   15

Law 3:   Energy Can Be Changed By Matter   25

Law 4:   Energy Moves Forward   37

Law 5:   Energy Can Move At Different Speeds   49

Law 6:   Energy Has Different Sizes And Shapes   75

Law 7:   Energy Has Different States   93

Law 8:   Atoms Are Energy Particles Traveling In Circles   103

Law 9:   An Element Is A Chain Of Proton Rings   123

Law 10: Molecules Are Elements Chained Together By Shared Electron Rings   177

Law 11: Heat / Cold Is Caused By The Size And Speed Of An Atom's Electron Rings   197

Law 12: All Molecules Have Energy Streams That Travel Through Their Centers   209

Law 13: Gravity Is An Energy Stream That Pushes All Molecules In Its Path — 217

Law 14: A Gravity Streams Length Can Reach As Far As Light — 231

Law 15: Sound Is A Particle In A Corkscrew Energy State — 239

Law 16: Light Is A Tadtron In 5 Different States — 259

Law 17: Electricity Is A Particle That Swims — 277

Law 18: Magnetic Fields Are Caused By Circling Energy Particles — 289

Law 19: Exiting Gravity Changes State After Passing Through The Atom — 335

Law 20: U = #Td Particle Mechanics - The New Science — 353

Practical Applications — 359

The Author's Source — 371

Preface

What is the most sought after thing in physics? The unifying theory is. What is the unifying theory? It's the answer to how the universe works. If they are looking for it, then they don't have it. This is an admission that physics does not currently know how the universe really works. Why can't physics produce this answer? Have you ever worked out a long mathematical equation? If you have one number wrong, you will not get the correct answer. If you have two numbers wrong, you will not get the correct answer. If you have several numbers wrong, you will not get the answer. How do you get the answer to a mathematical equation that has several numbers wrong? Fix all the wrong numbers, then rework the equation.

The universe is more like a jigsaw puzzle than an equation. This book will treat how the universe works like a jigsaw puzzle. How do you put a jigsaw puzzle together? First, you open the box. Second, you spread out the pieces on the table. Third, you turn all the pieces right side up. Fourth, you build the frame work. Fifth, you start putting sections of same color pieces together. Sixth, you put in the final pieces. Last, you look at the picture. What happens if some of the puzzle pieces are upside down? You will not be able to put the puzzle together. It will never work, no matter how hard you try. What happens if the framework isn't assembled correctly? You will never get the puzzle put together.

What's wrong with current physics? Several pieces of the puzzle are upside down? Which ones? First, atoms and molecules are not tiny solar systems. They have to be something different. They must provide the underlying mechanical workings of gravity. Second, gravity is not a force or an attraction that pulls. It is something completely different. Third, there is nothing behind the forces of nature. There are particles behind every type of energy. Fourth, waves are a three dimensional phenomenon. You can't treat them as two dimensional. Fifth, the universe is not moving to a state of disorder. Sixth, there is not a speed limit. The ideas and concepts above are based on 18th, 19th and 20th century ideas. If you can open your mind to the possibility that these things are wrong, then you will be able to move to the next level. If you are happy with all these things, don't waste your time reading this book.

This book approaches physics like a jigsaw puzzle. There are 20 laws that have been laid out. The first 7 laws lay the pieces of the puzzle out on the table. Many of the puzzle pieces are separated, grouped and turned right side up. Law 8 and 9 are the framework or outside edge of the puzzle. Law 10 thru 20 put the rest of the puzzle in place. Each of these laws deals with an individual section of physics. Anything that is upside down is turned right side up. When the puzzle is together, you can see a picture. The picture is a simple concept of how the universe works. Law 1 through 7 are very basic and very fundamental. Law 8 through 20 build on these and introduce all the neat new parts of the theory.

This book was written without all the hard to understand, scientific jargon. Common easy to understand words are used whenever possible. You will be introduced to many new terms, new ideas and new concepts. There are over a hundred full color illustrations. The illustrations and their descriptions are on opposing pages. You can see the illustrations as you read. Your mind will be opened to a new way of looking at energy and the universe. We will be going on a journey. We are going to follow a single particle as it travels across the universe. Our particle will change and perform many tasks. When you finish reading, you will have to answer a question. Is Particle Mechanics the Unifying Theory? You will have to be your own judge. Physics is riddled with theories. History is the judge of each theory. This is one of the first theories of this century. Welcome to 21st Century Physics!

Law 1

Law 1

Energy can move through matter

Law 1: Opening comments

In this law we will be showing how energy moves through matter. The different types of energies we will be looking at are light, radio, magnetic, gravity, electricity, X-ray, infrared light and sound. All these energies have characteristics that they can go through matter. Some of the matter is ordinary things we are familiar with. All these energies travel through the matter. Think of this as traveling through a door in a house. If the door is too small you can't get through. Think of each type of matter. What kind of door or opening does it have so that the energy can travel through?

This is a big part of the jigsaw puzzle. We are grouping pieces of the puzzle. These pieces have the same characteristics. Look at both things, the energy and the matter. Put yourself as the energy. Think of yourself as a little tadpole. You are swimming. You have eyes to see. Think about what you might see as you travel through the matter. Think about the doors or openings you go through.

I like to use illustrations, so you can see what I am talking about. We used a dog named Taddles in several of the illustrations.

Light can move through matter

In this observation the energy is light and the matter is glass. Take a flashlight and shine it on a piece of glass. The light will proceed through the glass and the light will shine on the other side. If the glass is clean and very clear you would not be able to tell the glass is even there. Not only is the light traveling through the glass, it is traveling through the air as well. Take a fish aquarium and shine a light behind it, you will have the same result. The light will travel through the air, through glass, through water, through glass, and then shine on the other side of the aquarium. In this experiment anyone can see and prove this. Energy can move through matter. Energy (light) can move through a gas, a liquid, or a solid.

Illustration 1-1: Light goes through glass

Light goes through glass

Glass

Energy moves through matter

Radio waves go through walls

In this observation the energy is a radio wave and the matter is the wall in my house. Somewhere there is a radio station. The radio station has a transmission tower. They turn the tower on and "radio waves" are sent out. I can be sitting in my house with all the doors shut. I can turn on the radio and the radio will pick up the "radio waves" that were sent from the radio stations transmission tower. Whatever is sent, music or a talk show, I can pick it up. The radio waves have passed through the wall of my house. The radio waves have traveled through the air too. If I were on a submarine, I could pickup the "radio waves" under water to a certain depth. Energy can move through matter. Energy (radio waves) can move through a gas, a liquid, or a solid.

Illustration 1-2: Radio waves go through walls

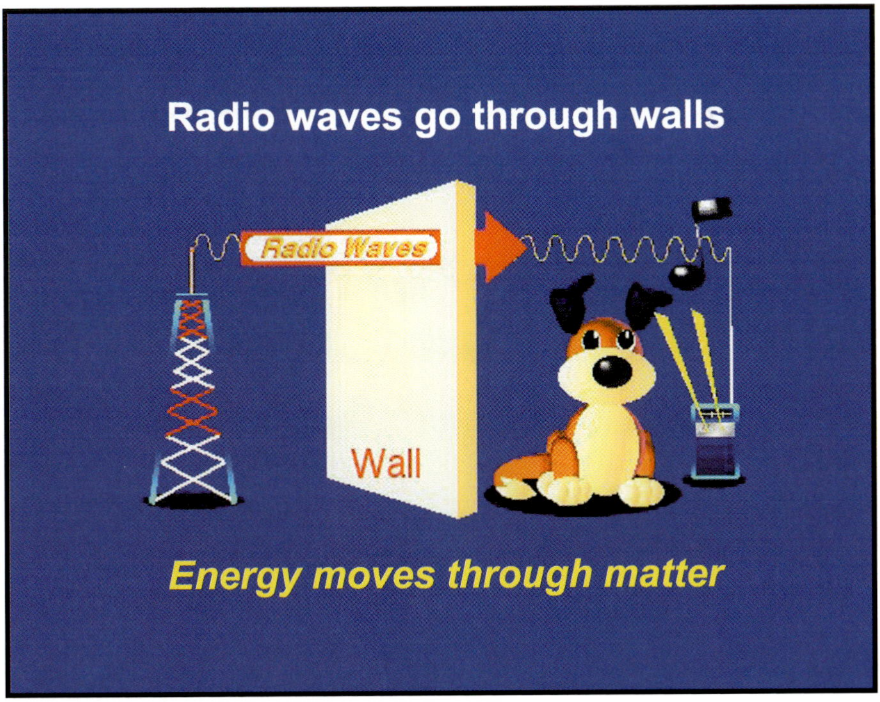

Magnetic waves go through cardboard

In this observation the energy is a magnetic field produced by a magnet and the matter is a piece of cardboard. Place a magnet on one side of a piece of the cardboard. Then place another magnet on the other side. They will act as if the cardboard is not there. The magnets will stick together or repel each other depending on the north and south poles. The magnetic field of both magnets passed through the matter of the cardboard. A piece of glass will act the same as the cardboard. Next take two pieces of glass. Put water in between the glass, a very thin aquarium. The magnets will still work the same. Magnets also work with just air between them. Energy can move through matter. Energy (magnetic field) can move through a gas, a liquid, or a solid.

Illustration 1-3: Magnetic waves go through cardboard

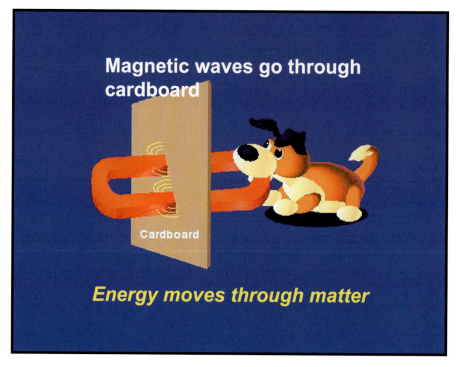

Gravity goes through matter

In my fourth observation the energy is the earth's "Gravitational Pull" and the matter is Taddles, a 20-pound dog. If Taddles jumps in the air, a force or energy causes him to return back to the earth. If Taddles stands on a scale, it will say Taddles weighs 20 pounds due to a force exerted on him. If you slide a 12-inch thick steel plate underneath the scale, it will still say Taddles weighs 20 pounds. The conclusion is that the " Gravitational Pull" goes through the air, through Taddles, through the scale and through the 12-inch steel plate. It does not matter how thick the steel plate is, the scale will still say Taddles weighs 20 pounds. If you suspend a twelve inch plate of steel above Taddles, he will still weigh 20 pounds. If you suspend a glass aquarium full of water above Taddles, he will still weigh 20 pounds. The gravity goes through the air, the glass, the water, Taddles and the scale. Energy can move through matter. Energy (gravity) can move through a gas, a liquid, or a solid.

Illustration 1-4: Gravity goes through matter

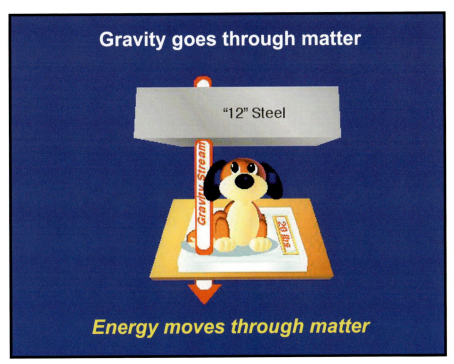

Electricity moves through a silver wire

In my fifth observation the energy is electricity and the matter is a piece of silver wire. When you put an electrical charge at one end of a piece of silver wire and you have a light bulb at the other end, the light bulb will give off light. The electricity moves through the silver wire. Another example is putting an electrical charge in a liquid. The electricity will travel through the liquid. Another example is lightning. It travels through the air. Energy moves through matter. Energy (electricity) can move through a gas, a liquid, or a solid.

Illustration 1-5: Electricity moves through a silver wire

Electricity moves through a silver wire

Silver Wire

Energy moves through matter

X-rays pass through skin

In my sixth observation the energy is x-rays and the matter is Taddles, the dog. A dog is made up of both solids and liquids. We will use one example of a solid and one example of a liquid. Taddles' bones are the solids. Taddles' blood is the liquid. To do this example you turn on the x-ray machine. It produces x-rays. The x-rays first travel through the air. Then they travel through Taddles. The x-rays then strike the "x-ray film" on the other side of Taddles. The x-rays act differently on what they go through. Most of the x-rays go through Taddles' blood and turn the film black on the other side. Not all the x-rays go through the bones, so the x-ray film appears a little lighter on the other side. When you look at the x-ray film, you will be able to see an outline of Taddles. You will also be able to see all his bones. Energy moves through matter. Energy (x-rays) can move through a gas, a liquid, or a solid.

Illustration 1-6: X-rays pass through skin

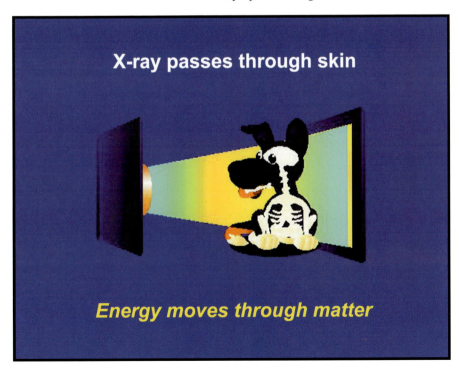

Infrared light goes through glass

In my seventh observation the energy is infrared light and the matter is glass. For our example we will use a TV and a remote control device. By pressing buttons on the remote control you can turn the TV on and off, turn the volume up and change the channels. If you put a piece of glass in front of the remote control, it will still work. Remote control devices send out infrared signals. The infrared light will proceed through the glass and control the TV. Not only is the infrared light traveling through the glass, it is traveling through the air as well. Take a glass of water and shine the infrared light through it. You will get the same results. The infrared light will travel through the air, through glass, through water, through glass, and control the TV. This experiment anyone can do and prove. Energy can move through matter. Energy (infrared light) can move through a gas, a liquid, or a solid.

Illustration 1-7: Infrared light goes through glass

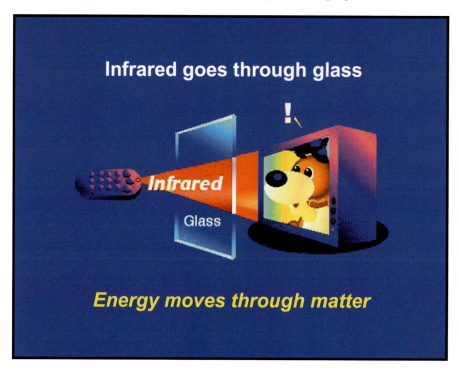

Sound goes through walls

In my eighth observation the energy is sound and the matter is a wall in a house. Outside the house is Taddles, the dog. All the doors and windows are shut. If Taddles barks, you will be able to hear him. The bark will make a sound. The sound will travel through the air and through the wall. Then the sound will travel to your ear. The sound will vibrate your ear drum. You will be able to "hear" the sound. I don't have a good example of filling the wall with water. So a different example would be whales underwater. They communicate with each other in the water in the ocean. Energy can move through matter. Energy (sound) can move through a gas, a liquid, or a solid.

Illustration 1-8: Sound goes through walls

Law 1: Closing comments

Did you come to any conclusions after reading this? There are a couple of things you can be sure of. One conclusion is that there are many doorways or openings in matter. There is space or openings that the energy can go through. A second conclusion it that the energy is small enough to go through the holes or openings.

We have grouped some of the pieces of the jigsaw puzzle. We can see a small piece of the big picture. Energy can move through matter.

Law 2

Law 2

Energy can be reflected by matter

Law 2: Opening comments

In this law we will be showing how energy can be reflected by matter. The different types of energies we will be looking at are light, radio, electricity, infrared light and sound. All these energies have characteristic that cause them to be reflected from matter. The matter in the next observations is ordinary things we are familiar with. All these energies bounce off of the matter. The energies are like a baseball bouncing off of a wall.

This is another big part of the jigsaw puzzle. We are grouping pieces of the puzzle. These pieces have the same characteristics. Look at both things, the energy and the matter. Be the energy. Think of yourself as a little tadpole. You are swimming. You have eyes to see. Think about what you might see as you collide with the matter.

Light is reflected from a mirror

In this observation the energy is light and the matter is a mirror. A mirror is nothing more than a piece of glass with a thin coat of silver. Shine a flashlight on the mirror. The light will bounce off or be reflected by the mirror. There may be as much as 97 % of the light reflected. Where does the other 3 % go? The light can be reflected at almost any angle from the mirror. Take a flashlight at night and shine it on a body of water. Some of the light will be reflected. If you are on a small lake, you can see the light reflected on the bank on the other side. For another example we could go to a desert in the day time. If you are in a flat area you can see water vapor evaporating. If the sun is at a right angle, some of the sunlight will be reflected off of the evaporating water. Energy can be reflected by matter. Energy (light) can be reflected by a gas, a liquid, or a solid.

Illustration 2-1: Light is reflected from a mirror

Matter reflects different amounts and kinds of light

In my second observation the energy is light and the matter is a tree. When light strikes the tree, some of the light is reflected and the rest is absorbed or changed. When we see something, it is the light reflected from an object coming to our eyes. Our eyes take this reflected light and paint a picture in our mind. The light that comes from the tree also has color with it. For example light strikes a leaf on the tree. The green light is reflected from the leaf and goes to your eyes. Your eyes then tell you the leaf is green. The rest of the light that struck the leaf was absorbed, changed, or disappears. Where did the rest of the light go? Another example is light striking the trunk of the tree. Some light is reflected to your eyes and you see brown. The light may be a combination of red light, green light and blue light. Your eyes paint a picture of brown. Shine a light on green paint. It will reflect green light back to your eyes. Take a smoke bomb that has sulfur in it. Light will be reflected by the smoke. The smoke will reflect all the yellow light and you will see yellow. Energy can be reflected by matter. Energy (light) can be reflected by a solid, a liquid, or a gas.

Illustration 2-2: Light is reflected from a leaf

Light is reflected from a leaf

Energy can be reflected by matter

Radio waves can be reflected by matter

In my third observation the energy is a radio wave and the matter is the upper atmosphere. There is a limit to how far radio waves will pass through matter. If you are deep in the earth your radio will not work. When you get to far from a radio tower your radio will not work because the radio wave must pass through the earth to get to you. But sometimes in the evening or night or early in the morning you can pick up a radio signal much farther away. Sometimes its thousands of miles away. Why? It's because the radio waves bounce off or are reflected by the upper atmosphere. These radio waves are then deflected down to your radio and you can receive the wave. Another example is having a radio at the bottom of a canyon. The radio tower can be behind you and the earth is too thick for the radio waves to go through. You can still hear the radio because the radio waves can be reflected off of the other canyon walls. I don't have a good example of a radio wave being reflected from a liquid. Energy (radio waves) can be reflected by matter. Radio waves can be reflected by a solid or a gas.

Illustration 2-3: Radio waves reflect off of the atmosphere

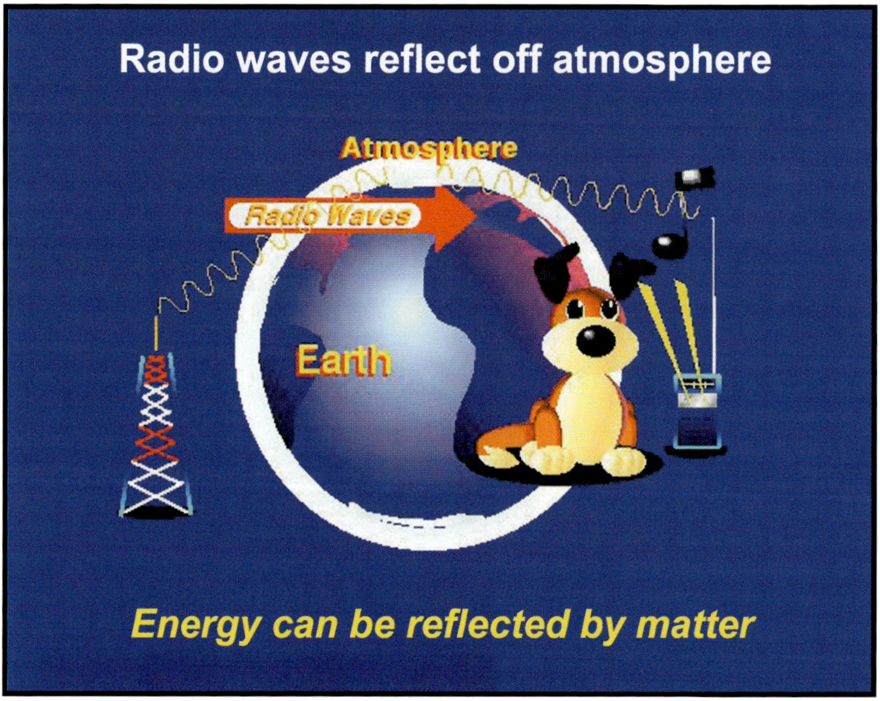

Energy can be reflected by matter

In my fourth observation the energy is electricity and the matter is glass. To do this observation, we will take a piece of silver wire and plug it into an electrical outlet. If you touch the silver wire to a light bulb, the light bulb will produce light. The electricity will travel through the wire. If you touch the electric wire to a piece of glass, nothing will happen. The electricity doesn't come out of the wire, it is reflected by the glass. If you touch the electric wire to a glass full of pure water, nothing will happen. The electricity will be reflected by the water. (Pure water doesn't conduct electricity). If we hold the wire in the air, nothing will happen. The air reflects the electricity. Energy (electricity) can be reflected by matter. Electricity can be reflected by a gas, a solid or a liquid.

Don't try this experiment, a slight impurity in water will cause it to carry an electrical current.

Illustration 2-4: Electricity is reflected from glass

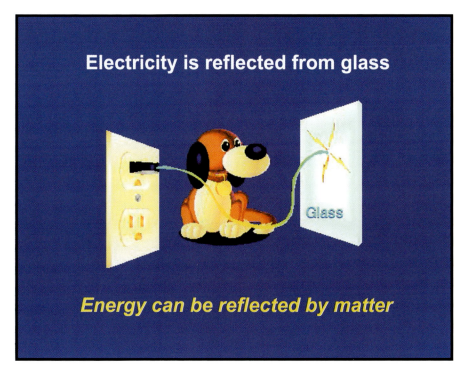

Energy can be reflected by matter

In this observation the energy is infrared light and the matter is a mirror. To do this example we need a mirror and a TV remote control device. Remote control devices send out an infrared signal that is used to control the TV with. You can use the remote control to turn the TV on and off, change the channels and change the volume. Hold the mirror behind you. Shine the remote control at the mirror. You can control the TV just the same as if you were pointing the remote control directly at the TV. Put a mirror on the ceiling, point the remote control at the ceiling, it will control the TV. I don't have an example of infrared light reflecting off of a gas or a liquid. Energy (infrared light) can be reflected by matter. Energy (infrared light) can be reflected by a solid.

Illustration 2-5: Infrared is reflected from a ceiling

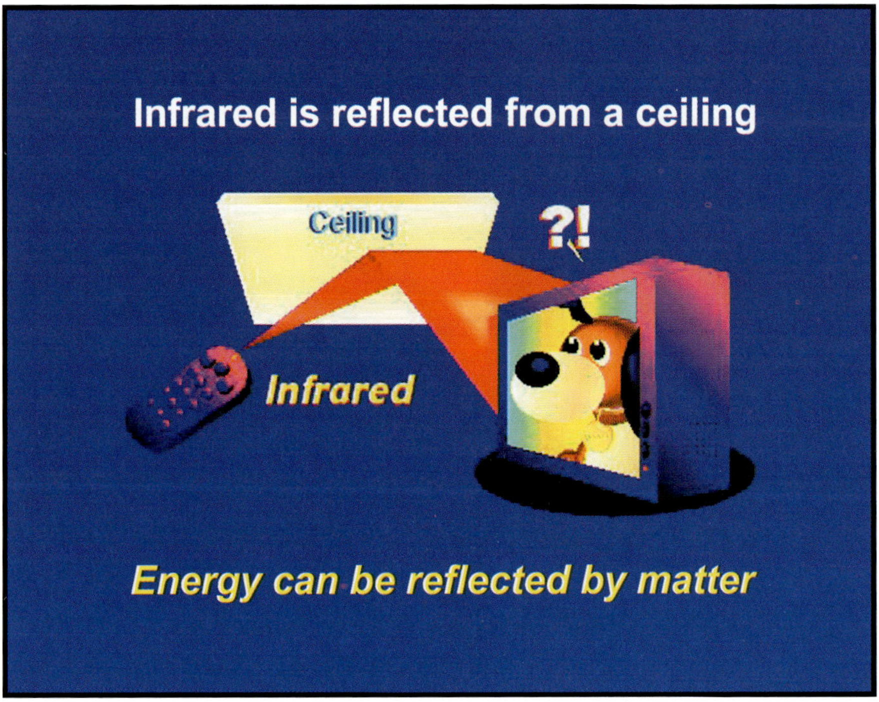

Energy can be reflected by matter

In this observation the energy is sound and the matter is a wall in a canyon. To do this example we have Taddles, the dog, standing in a canyon barking. When Taddles barks, the sound will reflect off of the canyon walls. This is called an echo. The echo will sound like the bark but it won't be as loud. Sound will echo off of water too. Anyone can go out on a boat and talk. They will be heard much better because the sound will be reflected up off of the water. I was out on a river one day fishing with a friend. We were talking in a normal tone of voice and my wife could understand what we were saying almost a mile away. I don't have an example of sound being reflected by a gas. Energy (sound) can be reflected by matter. Energy (sound) can be reflected by a solid or a liquid.

Illustration 2-6: Sound echoes from a canyon

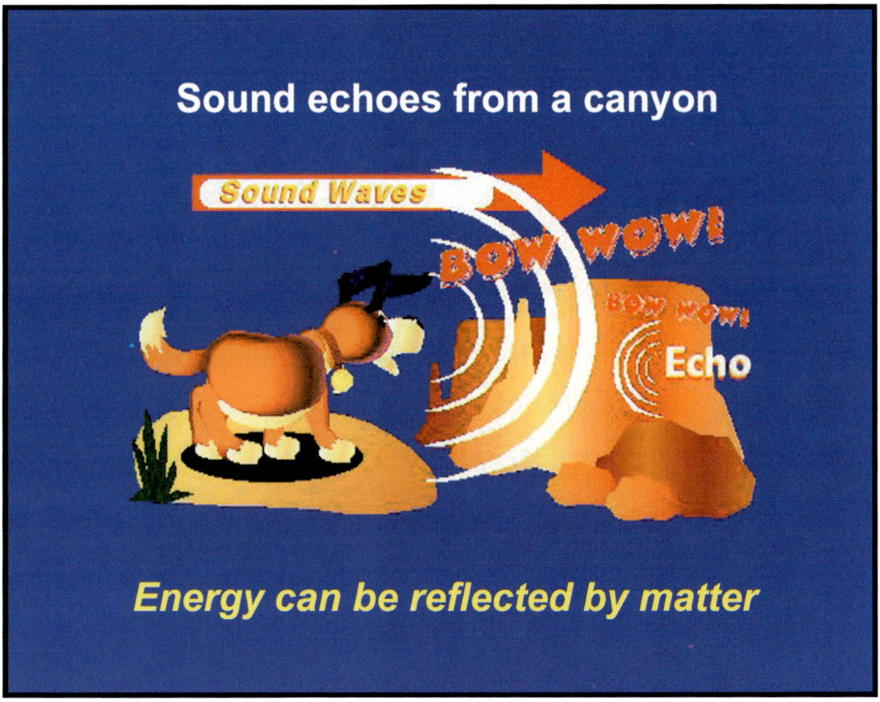

Law 2: Closing comments

If we throw a baseball against a wall, what happens? It bounces off. It doesn't bounce back as fast. The baseball loses a lot of it's speed. The energies don't. The light comes in at the speed of light and it's reflected at that same speed. Sound is reflected at the speed of sound. Energies can change directions without any loss of speed. The matter has different shapes and sizes. That's why different energies bounce off differently. In law 1 we saw how there were openings in the matter. In this law we can see that there are spaces that energy can't go through.

We have grouped some of the pieces of the jigsaw puzzle. We can see another piece of the big picture. Energy can be reflected by matter.

Law 3

Law 3

Energy can be changed by matter

Law 3: Opening comments

In this law we will be showing how energy can be changed by matter. The different types of energies we will be looking at are light, radio, electricity, magnetic, x-ray, and sound. All these energies have characteristic that cause them to be changed by matter. The matter in the next observations are ordinary things we are familiar with. All these energies will be changed when they come in contact with matter. The energies are like an egg. When you throw an egg against the wall, it changes.

This is another big part of the jigsaw puzzle. We are grouping pieces of the puzzle. These pieces have the same characteristics. Look at both things, the energy and the matter. Imagine yourself as the energy. Think of yourself as a little tadpole. You are swimming. You have eyes to see. Think about what you might see as you collide with the matter. This time you are going to change when you hit the matter. Think of it this way. You go to a pond. You see hundreds of tadpoles swimming in the water. You come back a few weeks later. You see hundreds of frogs. You don't see any tadpoles. What happened?

Light can be changed by matter

In my first observation the energy is light and the matter is a piece of glass painted with flat black paint. Let the paint dry. Shine a light on the glass. There may be as much as 97 % of the light that is absorbed or changed. Only 3 % will be reflected. With the 3 % that is deflected, you can see the black color. But where did the other 97 % go? It's not light anymore. So where is it? If it was light and now it isn't, then it has changed. In the next example get some flat black paint before it dries. Then shine a light on it. You will get almost the same result as the black paint on the glass. Take an old tire and burn it. You will get some black smoke. Shine a light on the smoke and only a small amount will be reflected back. The rest of the light will be changed. Energy can be changed by matter. Energy (light) can be changed by a solid, a liquid, or a gas.

Illustration 3-1: Light can be changed by matter

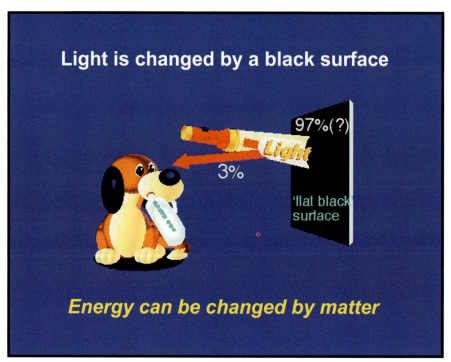

Radio waves can be changed by matter

In my second observation the energy is a radio wave and the matter is a radio. The radio wave comes into your radio. You hear the music, sound comes out. Where did the radio wave go? You can't replay the music. It was a radio wave and now it isn't. It has changed. If you are deep in a cave, a radio won't work. It won't work because the radio waves are changed by the solid material in the earth before they get to the radio. If you are at the bottom of the ocean, a radio won't work. It won't work because the radio waves are changed by the liquid in the ocean before they get to the radio. Have you ever listened to a radio when a storm came in? You may be getting perfect reception. Then when the storm comes in you will get crackles and pops and then you can't hear anything. You lose reception. Why? The thicker clouds have changed the radio waves. Energy can be changed by matter. Energy (radio waves) can be changed by a solid, a liquid or a gas.

Illustration 3-2: Radio waves can be changed by matter

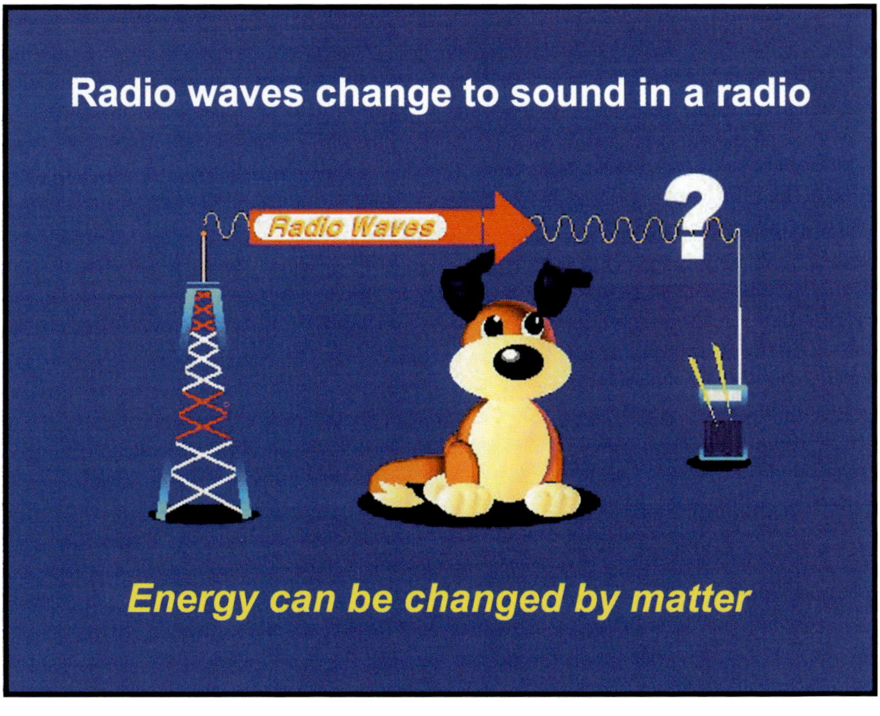

Electricity can be changed by matter

In my third observation the energy is electricity and the matter is a piece of silver wire. When you put an electrical charge at one end of a piece of silver wire and you have a light bulb at the other end, the light bulb will give off light. The electricity moves through the silver wire. The current can be measured as it travels through the wire. After the electricity goes to the light bulb, it will be gone. Where did the electricity go? Where did the light come from? There was some electricity now it's gone. It has changed. Stick the same wire into a pond of water. It will shock the fish. The electricity goes into the water and is gone. The electricity has changed. Run electricity into a neon light. The neon light will give off light and the electricity is gone. Energy can be changed by matter. Energy (electricity) can be changed by a solid, liquid, or a gas.

Illustration 3-3: Electricity can be changed by matter

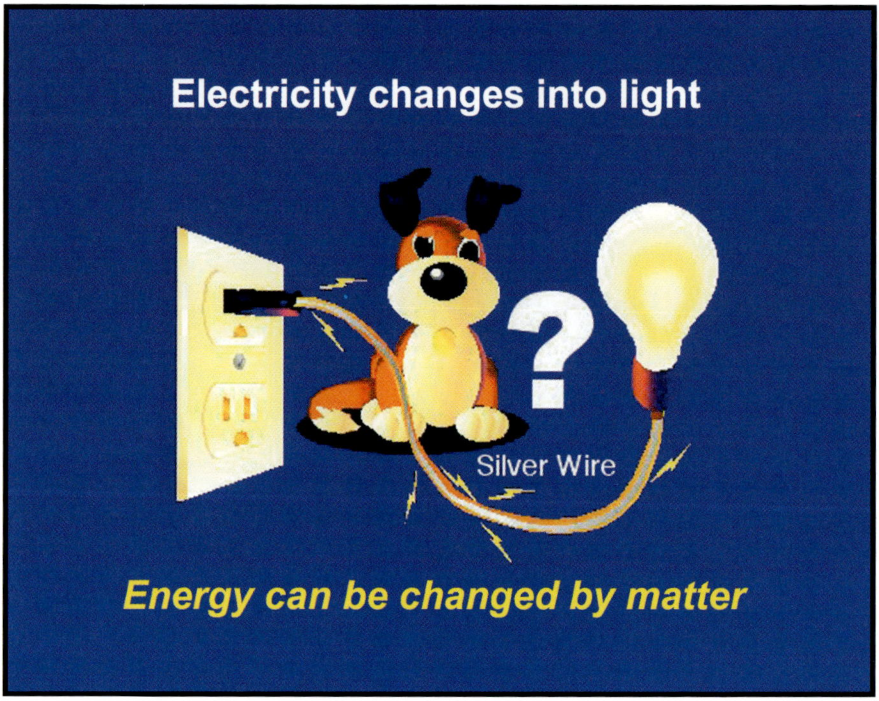

Magnetic energy can be changed by matter

In my fourth observation the energy is a magnetic field and the matter is an electrical generator. When you pass a magnet by a coil of wire, a current of electricity is produced. This process is called a generator. Supposedly the electricity comes from electrons out of the magnet field and is then changed to electricity. You can run a generator for years and these electrons are never depleted. According to what I know about the current laws of physics, this electron source should soon become depleted. So where did the electrons come from? It was magnetic energy now it is electricity. It has changed. Energy (magnetic) can be changed by matter.

Illustration 3-4: Magnetic energy can be changed by matter

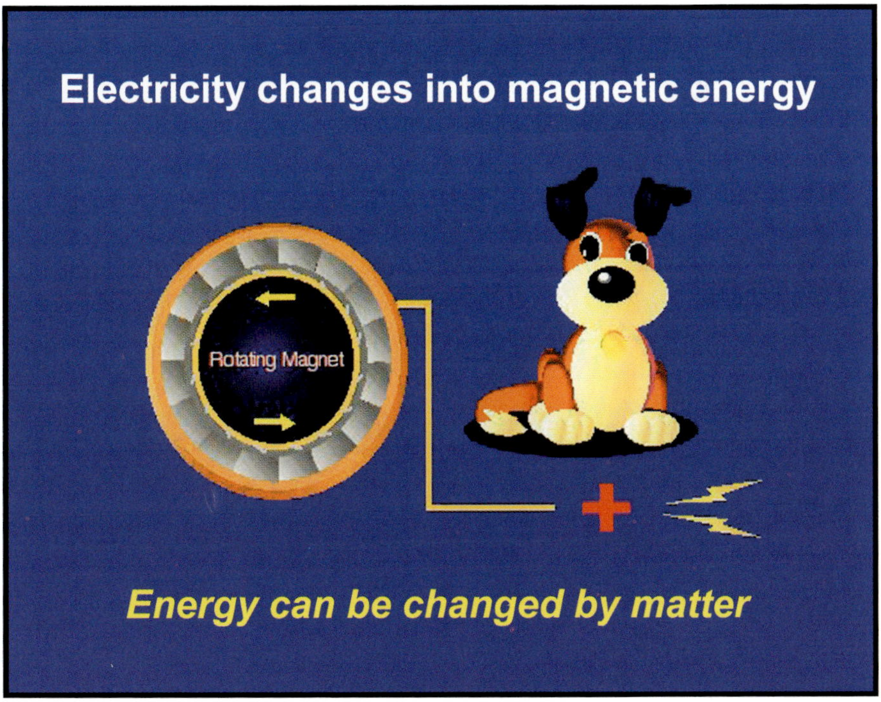

Electricity can be changed by matter

In my fifth observation the energy is electricity and the matter is an electromagnet. To make an electromagnet you can take a wire and wrap it around an iron nail. Next run electricity through the wire around the nail. It is now an electromagnet. Now place the electromagnet next to another iron nail. It will pick up the iron nail just like a magnet would. The electricity that was in the wire that was circling around the nail was changed to magnetic energy. Now turn off the electricity. The iron nail will fall. Where did the magnetic energy go? It was electricity then it was magnetic energy. Now it's gone. It has changed. Energy (electricity) can be changed by matter.

Illustration 3-5: Electricity can be changed by matter

X-rays can be changed by matter

In my sixth observation the energy is an x-ray and the matter is a skeleton. Take an x-ray of Taddles. Most of what you will see on the x-ray film is Taddles' skeleton. Each of the dog's bones can be seen. The x-ray film records what x-rays strike the film. Where the bones are seen, are places where the x-ray did not reach the film. The x-rays did not go through the bones. They were x-rays. Now they are gone. They have changed. Where did they go? The x-rays that strike the film are gone. They changed too. Where did they go? Energy (x-rays) can be changed by matter.

Illustration 3-6: X-rays can be changed by matter

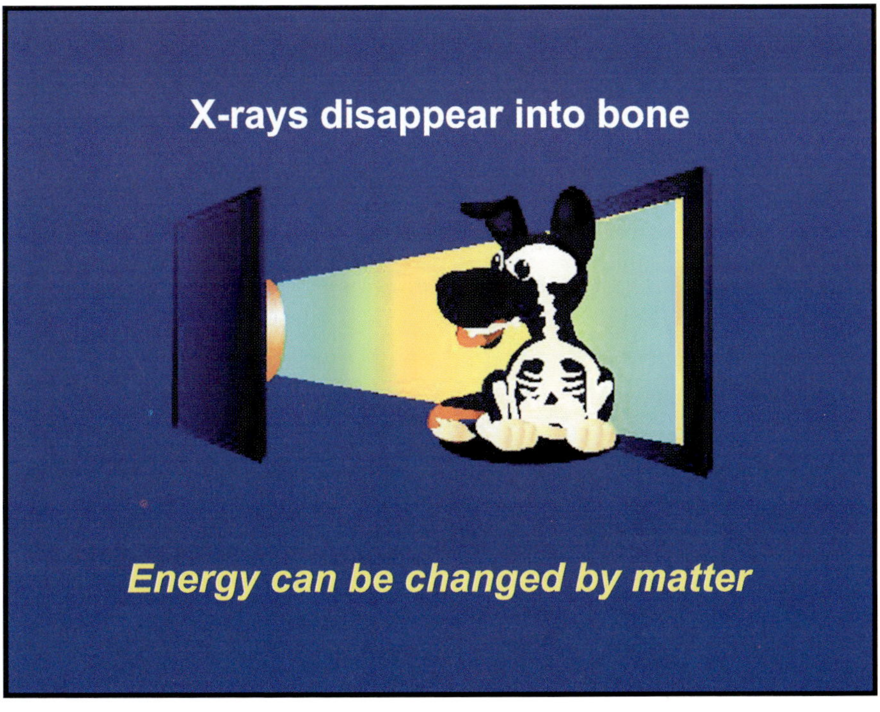

Sound can be changed by matter

In my seventh observation the energy is sound and the matter is a soundproof room. Put Taddles inside the soundproof room. Taddles will now bark. The sound will strike the soundproof walls. If you are standing on the outside of the soundproof room, you will hear nothing. The sound hits the walls and now it is gone. Where did the sound go? Energy (sound) can be changed by matter.

Illustration 3-7: Sound can be changed by matter

Law 3: Closing comments

If we throw an egg against the wall, what happens? It breaks and splatters. It's not an egg. When we went back to the pond and saw frogs, is that a clue to where the tadpoles went? What if we went back to the pond and there were hundreds of frogs and hundreds of tadpoles? What would we think? The frogs magically appeared and the tadpoles stayed the same. No, we would know that the tadpoles changed to frogs and some new tadpoles hatched out.

Energy is constantly changing. It can be something that we can measure or feel or hear or see or touch. But what about the energy we cannot measure or feel or hear or see or touch? There are some "unknown energies" out there. Energy can change very quickly, maybe even instantly. Just because we can't see the change doesn't mean it isn't any different than a tadpole changing into a frog.

We have grouped some more of the pieces of the jigsaw puzzle. We can see another piece of the big picture. Energy can be changed by matter.

Law 4

Law 4

Energy moves forward

Law 4: Opening comments

In this law we will be showing how energy moves forward. The different types of energies we will be looking at are light, radio, gravity, electricity, x-ray, sound and magnetic. All these energies have characteristic that cause them to move forward.

This is a small part of the jigsaw puzzle. We are grouping pieces of the puzzle. These pieces have the same characteristic of moving forward. They don't stop. They go from one destination to the next. Think of a pond full of tadpoles. They are all swimming. They all swim forward.

Light moves forward

In my first observation the energy is light and the matter is glass. Shine a light and it will proceed through the glass and the light will shine on the other side. In this example the light moves forward through the glass. If it were a mirror and it was reflected it would continue forward on until it meets another object. Light from the sun moves millions of miles forward from the sun to the earth. Energy (Light) moves forward.

Illustration 4-1: Light moves forward

Radio waves move forward

In my second observation the energy is a radio wave and the matter is the walls of a house. First a radio station produces a radio wave. It sends out radio waves in all directions. Taddles, the dog, is sitting inside the house. You turn the radio on and listen. Taddles can hear the music. The radio waves have passed straight through the walls. The radio waves have traveled forward from the radio station through the air, through the walls to the radio. Energy (radio waves) moves forward.

Illustration 4-2: Radio waves move forward

Gravity moves forward

    In my third observation the energy is the earth's "Gravitational Pull" and the matter is Taddles. If Taddles jumps in the air, a force or energy causes Taddles to return back to the earth. The "Gravitational Pull" always "pulls" straight back to the center of the earth. The gravity is always moving forward to the center of gravity of the larger object. The gravity is moving forward through all things. Energy (gravity) moves forward.

Illustration 4-3: Gravity moves forward

Electricity moves forward

    In my fourth observation the energy is electricity and the matter is a piece of silver wire. When you put an electrical charge at one end of a piece of silver wire and you have a light bulb at the other end, the light bulb will give off light. The electricity moves forward through the silver wire to the light bulb. It doesn't matter if the wire is straight or crooked, it will go forward until it reaches the light bulb. Energy (electricity) moves forward.

Illustration 4-4: Electricity moves forward

X-rays move forward

In my fifth observation the energy is x-rays and the matter is Taddles. The x-rays are generated from the x-ray machine. The x-rays move forward through Taddles and hit the x-ray film. Energy (x-rays) move forward.

Illustration 4-5: X-rays move forward

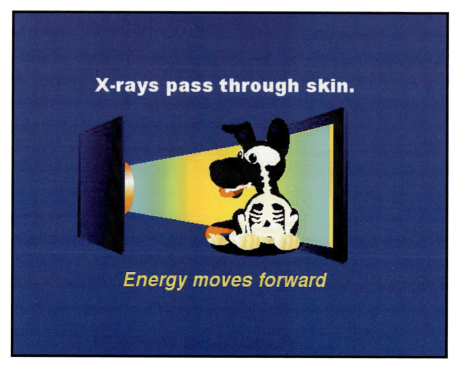

Sound travels forward

In my sixth observation the energy is sound and the matter is the air. If Taddles barks, the sound travels forward through the air. The sound will travel forward until it strikes something. If it hits the canyon wall and is reflected, that sound will travel forward until it hits something and is changed. Energy (sound) moves forward.

Illustration 4-6: Sound travels forward

Magnetic energy moves forward

    In my seventh observation I will use magnets. Magnets are an unusual part of nature. They "push" and "pull" each other. They "pull" objects made of iron. In the observation we are going to observe the way the "Magnetic Force Lines" travel. If you take 2 magnets and push 2 north poles together, you can feel a "circle" of these "Magnetic Force Lines". If you put iron filling around a magnet, you can see these force lines moving in a circle. There is a magnetic viewing film, which you can observe this "Energy" moving in a circle. I will explain later what I think this "Magnetic Force Line" is. It appears that magnets produce energy that moves forward in a circle. Energy (magnetic field) moves forward.

Illustration 4-7: Magnetic energy moves forward

Magnetic energy moves forward in a circle -

Energy (magnetic field) moves forward

Page 44

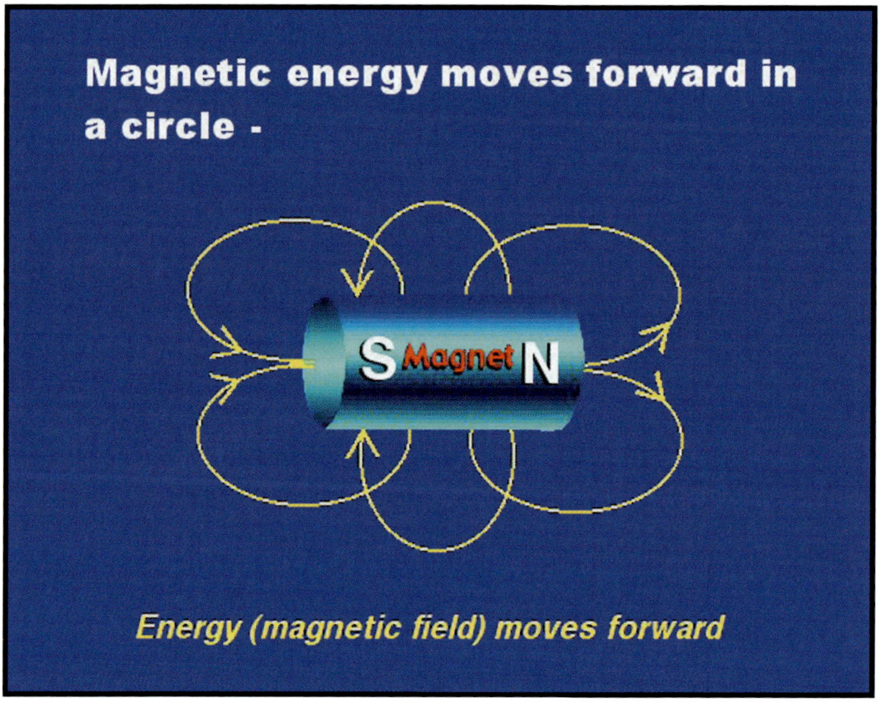

Law 4: Closing comments

This was not the most exciting part of the puzzle. Some pieces of a puzzle take time to put together. Those pieces are kind of like filler. You have to have them to see the final picture. Energy doesn't sit around. It moves forward. It has somewhere to go. It goes there. When energy reaches its destination it either goes on through, is reflected, or is changed.

Think of a pond full of tadpoles. They swim around. If they bump into something, they change their course and they swim another direction. Think of the tadpoles as always swimming. The tadpoles swim until they change into frogs. There are also different kinds of tadpoles. There are tadpoles that change into little brown frogs. There are tadpoles that change into little green frogs. There are tadpoles that change into bull frogs. All the different kinds of tadpoles swim differently. After the tadpoles change into frogs, they swim differently. They all can still move forward. As we go on, you will be able to see how the universe looks a lot like a pond full of tadpoles.

Law 5

Law 5

Energy can move at different speeds

Law 5: Opening comments

Energy is everywhere. It travels to and fro. It travels at different speeds. Those speeds can vary depending on the matter that the energies interact with. People have always put speed limits on how fast we can travel. When trains were going to travel at 30 mph, some people thought you would not be able to breath. Then we weren't going to travel faster than 100 mph. Then it was 200 mph. Then we weren't going to be able to go faster than the speed of sound. As we all know, all those speed limits have been broken. There is a new speed limit that can't be broken. That speed limit is the speed of light. Nothing can go faster than the speed of light, or so it has been said. I don't believe this. I believe light can go faster than the speed of light. I believe light can travel slower than the speed of light. In this law we will see that it is all in how you measure the speed.

At the turn of the 20th century, Albert Einstein came out with the theory of relativity. Einstein's whole theory was based on an idea that nothing could go faster than the speed of light. If anything can go faster than the speed of light, there are flaws in his theory.

This is another piece of the jigsaw puzzle. The speed at which energy travels is very relative to figuring out how things work. When you finish reading this law, you will know how to make light travel faster than the speed of light. You will know how to make light travel slower than the speed of light.

A grouping of energies

In this observation the energies are light, infrared light, electricity, radio waves, sound waves, x-rays, magnetic energy and gravity. I don't have the equipment to measure the speed of any of these energies. Scientists have the equipment and have put in the time to measure the speed of some of these energies. Light, infrared light, electricity, radio waves and x-rays all travel at about 186,262 miles per second(670,579,200 miles per hour). That's known as the speed of light. This speed is relative to the matter that they exit from. For example if you turn on a flashlight, the light will go away from the flashlight at 186,262 miles per second. The speed of sound is about 1,100 feet per second (750 miles per hour). The speed of sound changes with the type of material it is in. If it is traveling through water, it is 4 times faster than through the air. If sound is traveling through aluminum, it is 15 times faster than through the air. Sound is also relative to matter it is going away from. The speed of magnetic energy and gravity are unknown. Energy can move at different speeds. There is energy in the waves of water. Waves travel at different speeds.

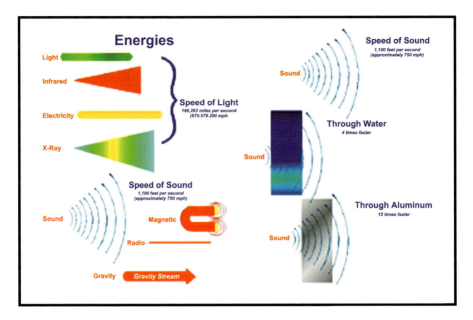

Illustration 5-1: A grouping of energies

Speed of bus plus speed of dog

In this example I am not going to show any energies. I will use two different types of matter. The first matter is Taddles, the dog. The second matter is a bus. We are going to measure speed. What is speed? Speed is the distance something can go in a measure of time. Simply put, it is how long it takes for something to get from point A to point B. I am going to convert all my examples into miles per hour (mph), so they will be easier to understand. In the next illustration we have a bus going 20 mph. We have point R at the rear of the bus (R is for rear). Taddles is going to run at 20 mph from point R on the bus to point F (the front of the bus). We only need one stop watch. We will time Taddles until he crosses point F. Taddles speed on the bus was 20 mph. On the ground, we have point A and point B. Point A is where Taddles starts. Point B is where Taddles is when he gets to point F. The distance is twice as far as R to F. Taddles ground speed was twice that of the speed on the bus. Taddles ground speed is the speed of the bus plus the speed of Taddles, which totals 40 mph. It takes the same amount of time to get from point R to point F as it does to get from point A to point B. If Taddles jumps off the bus while he is running and hits a wall, it will be the same effect as if Taddles was running at 40 mph.

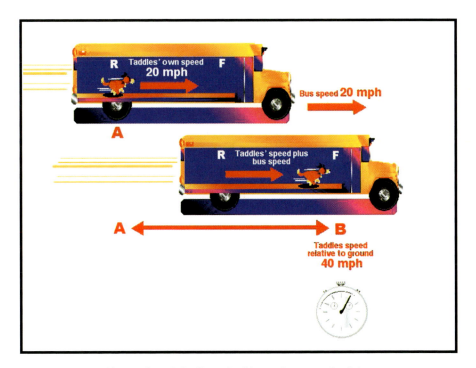

Illustration 5-2: Speed of bus plus speed of dog

Speed of dog minus speed of bus

In the illustration we have a bus going 20 mph. Taddles is going to run at 20 mph from the front of the bus to the rear of the bus. We only need one stop watch. We will time Taddles from point F to point R. His speed on the bus is 20 mph. We have point A under the bus on the ground. This is where Taddles starts to run. Point B is where Taddles will be when he gets to the rear of the bus. There is no distance between point A and point B. What was the ground speed of Taddles? It is the speed of Taddles minus the speed of the bus, which is 0 mph. It takes the same amount of time to get from point F to point R as is does to get from point A to point B. If Taddles jumps off the bus while running and hits a wall, it will be the same effect as if Taddles was standing still. It would be like running on a treadmill. You can run a mile on the treadmill and never go anywhere.

Illustration 5-3: Speed of dog minus speed of bus

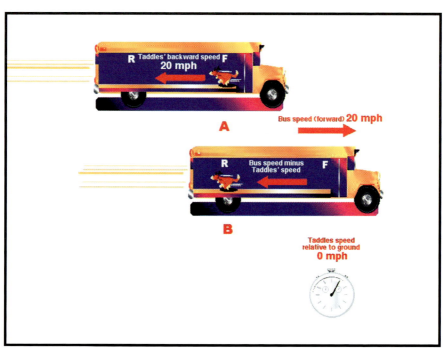

Speed of sound plus speed of bus

In this example sound is the energy. The matter is a bus. The bus is going 20 mph. Taddles is at the rear of the bus. He is going to bark from the rear of the bus. We only need one stop watch. We will time how long it takes the sound from the bark to get to the front of the bus. The distance is point R to point F. What is the speed? It is the speed of sound, which is about 750 mph (this figure is close and what we will use in this chapter). On the ground we marked point A, which is where Taddles barked. Point B is where the sound from the bark ended at the front of the bus. The distance from Point A to point B is a little more than the distance of point R to point F. It takes the same amount of time to cover both distances, which makes the ground speed of the bark faster. What is the ground speed of Taddles' bark? It is 750 mph plus 20 mph, which is 770 mph.

Illustration 5-4: Speed of sound plus speed of bus

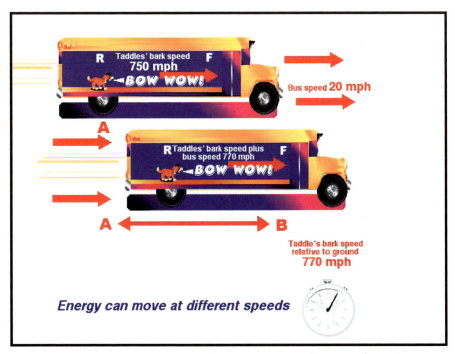

Speed of sound minus speed of bus

In this example sound is the energy. The matter is a bus. The bus is going 20 mph. Taddles is at the front of the bus. He is going to bark from the front of the bus. We only need one stop watch. We will time how long it takes the sound of the bark to get to the rear of the bus. The distance is point F to point R. What is the speed? It is the speed of sound, which is about 750 mph. On the ground we marked point A, which is where Taddles barked. Point B is where the sound from the bark ended at the rear of the bus. The distance from point A to point B is a little less than the distance of point F to point R. It takes the same amount of time to cover both distances, which makes the ground speed of the bark slower. What is the ground speed of Taddles' bark? It is 750 mph minus 20 mph, which is 730 mph.

Illustration 5-5: Speed of sound minus speed of bus

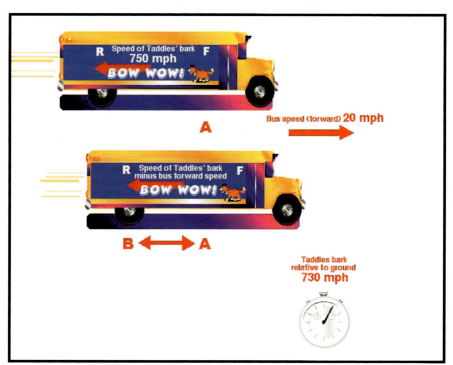

Speed of light plus speed of bus

    In this example light is the energy. The matter is a bus. The bus is going 20 mph. Taddles is at the rear of the bus. He is going to shine a flashlight from the rear of the bus to the front. We only need one stopwatch. We will time how long it takes the light from the flashlight to get to the front of the bus. The distance is point R to point F. What is the speed? It is the speed of light, which is about 670,579,200 mph. On the ground we marked point A, which is where Taddles turns on the flashlight. Point B is where the light from the flashlight ended at the front of the bus. The distance from Point A to point B is slightly more than the distance of point R to point F. It takes the same amount of time to cover both distances, which makes the ground speed of the light faster. What is the ground speed of Taddles' flashlights light? It is 670,579,200 mph plus 20 mph, which is 670,579,220 mph.

Illustration 5-6: Speed of light plus speed of bus

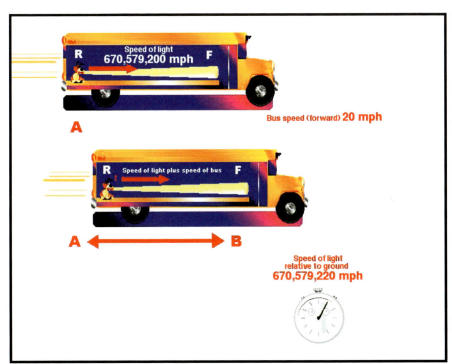

Speed of light minus speed of bus

In this example light is the energy. The matter is the bus. The bus is going 20 mph. Taddles is at the front of the bus. He is going to shine a flashlight from the front of the bus to the rear of the bus. We only need one stopwatch. We will time how long it takes the light from the flashlight to get to the rear of the bus. The distance is point F to point R. What is the speed? It is the speed of light, which is 670,579,200 mph. On the ground we marked point A, which is where Taddles turned on the flashlight. Point B is where the light from the flashlight ended at the rear of the bus. The distance from point A to point B is slightly less than the distance of point F to point R. It takes the same amount of time to cover both distances, which makes the ground speed of the light from the flashlight slower. What is the ground speed of Taddles' flashlights light? It is 670,579,200 mph minus 20 mph, which is 670,579,180 mph.

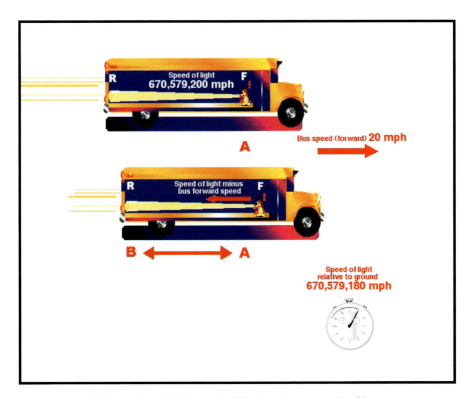

Illustration 5-7: Speed of light minus speed of bus

Speed of sound plus speed of sound

Up to this point we have used a bus traveling at 20 mph. The point A to point B was the ground speed of the matter or the energy. From here on I am going to refer to ground speed and not use Point A to Point B. We are now using a jet airplane that is traveling at the speed of sound.

In the next illustration, we have Taddles at the rear of a jet airplane. The airplane is traveling at the speed of sound. Taddles barks. The bark goes from the rear of the airplane to the front. The speed of the bark on the airplane is the speed of sound. What is the ground speed of the bark? It is the speed of the airplane plus the speed of the bark. It's the speed of sound plus the speed of sound, which is 1500 mph.

Now imagine a major league baseball pitcher at the back of the airplane. He throws a baseball to the front of the airplane at 100 mph. What is the ground speed of the baseball? It's the speed of the airplane plus the speed of the baseball. The ground speed of the baseball is 850 mph. It's to fast for me to hit.

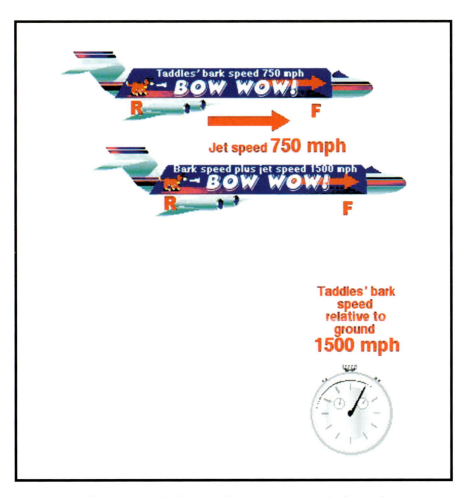

Illustration 5-8: Speed of sound plus speed of sound

Speed of sound minus speed of sound

In the next illustration, we have Taddles at the front of a jet airplane. The airplane is traveling at the speed of sound. Taddles barks. The bark travels to the rear of the airplane. The speed of the bark on the airplane is the speed of sound. What is the ground speed of the bark? It is the speed of the airplane minus the speed of the bark. It's the speed of sound minus the speed of sound, which is 0 mph.

Now imagine a major league baseball pitcher at the front of the airplane. He throws a baseball to the rear of the airplane at 100 mph. What is the ground speed of the baseball? It's the speed of the airplane minus the speed of the baseball. The ground speed of the baseball is 650 mph. It's still to fast for me to hit.

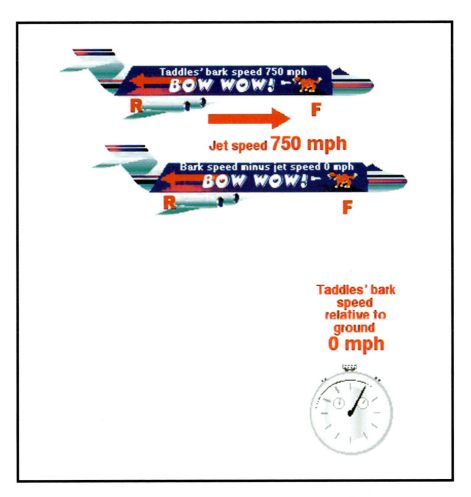

Illustration 5-9: Speed of sound minus speed of sound

Speed of sound plus speed of light

In the next illustration, we have Taddles at the rear of a jet airplane. The airplane is traveling at the speed of sound. Taddles shines a flashlight to the front of the airplane. The speed of that light on the airplane is the speed of light. What is the ground speed of that light? It is the speed of the airplane plus the speed of light, which is 670,579,950 mph.

This is why we went to all this trouble. We have shown that it is possible for something to travel faster than the speed of light. What traveled faster than light? Light did. That means light has a variable speed. Did time change? No, it didn't. We used only one stop watch to measure the speed. It comes down to how and where the speed is measured.

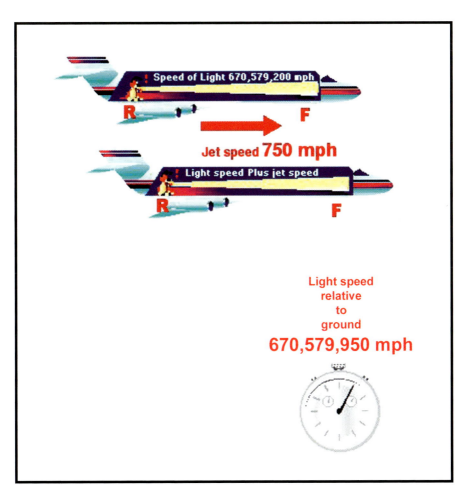

Illustration 5-10: Speed of sound plus speed of light

Speed of sound minus speed of light

In the next illustration, we have Taddles at the front of a jet airplane. The airplane is traveling at the speed of sound. Taddles shines a flashlight to the rear of the airplane. The speed of the light on the airplane is the speed of light. What is the ground speed of that light? It is the speed of the light minus the speed of the airplane, which is 670,578,450 mph.

In the last illustration, we showed it was possible to make light travel faster than the speed of light. In this illustration we have shown that it is possible for light to travel slower than the speed of light. Again it all comes down to how and where the speed is measured. Time did not change. Light has a variable speed. Let me add another thought. The earth is traveling though space in orbit around the sun. What is the sun surface speed of the light from the flashlight? Put in a point A and a point B and time it. You will have a different speed of light.

Illustration 5-11: Speed of sound minus speed of light

Speed of light plus speed of light

Everything we have shown so far, we have the capability to make and do. The spaceship in the illustration is theoretical. It is a spaceship that is traveling at the speed of light. I believe travel at this speed is achievable. How will light act on the ship? It won't be any different. Taddles is at the rear of the spaceship. He shines the flashlight to the front of the spaceship. The light will travel at the speed of light on the spaceship. The ground speed of the light will be the speed of light plus the speed of light, which is 1,341,158,400 mph.

In this example light is traveling 2 times the speed of light. It all depends on how and where you measure the speed. Let's give you another thought. If you were on that spaceship and you looked behind you, what would you see? You would not be able to see any light that was traveling slower than the speed of the ship. If there is some light going faster than the speed of the spaceship you would be able to see it. Behind the spaceship would look mostly black. Because most light is slower, or the same speed, than the speed of the spaceship.

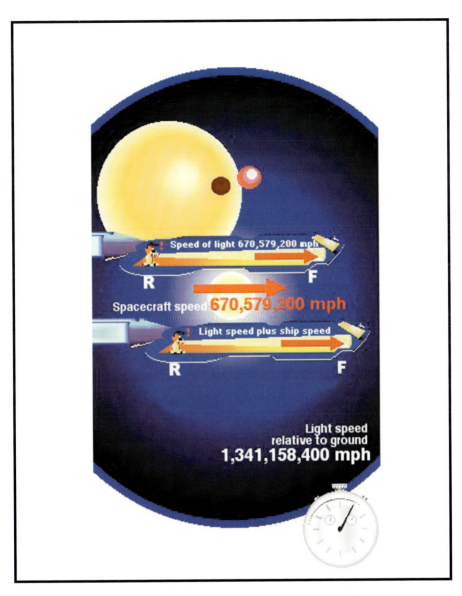

Illustration 5-12: Speed of light plus speed of light

Speed of light minus speed of light

We are going to use the same spaceship in this illustration. The spaceship is traveling at the speed of light. Taddles is at the front of the spaceship. He shines the light to the back of the spaceship. What is the speed of the light on the spaceship? It is the speed of light. What is the ground speed of the light? It is the speed of light minus the speed of the spaceship, which is 0 mph. The light is just sitting there and not moving, from the perspective of the ground. Once again the speed is based on where the speed is measured from.

In this example light is traveling at 0 mph. Would you be able to see it? Only if it came in contact with your eyes. If you where on the earth you would have to move to it to see it.

Let's give you another thought. If you were on the spaceship and you looked ahead of you, what would you see? You would be able to see all lights coming to you. The lights would appear to be twice as bright in comparison to a ship that is not moving.

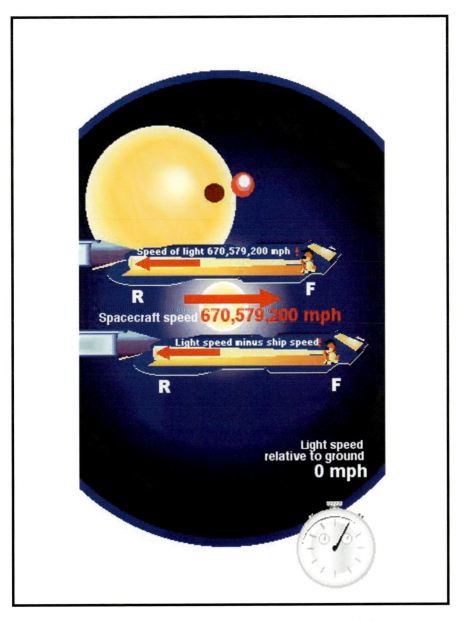

Illustration 5-13: Speed of light minus speed of light

Law 5: Closing comments

In this law we have seen how energies move at different speeds. We have seen how the same energy can travel at different speeds. We have seen that speed is relative to where it is measured from. If electricity is moving from the front to the back of the spaceship, what is its speed? If electricity is moving from the back to the front of the spaceship, what is its speed? The answers depend on where you measure the speed from. The same principle that we have seen in this chapter will work on all energies.

If light can go 2 times the speed of light, then Albert Einstein's theory of relativity is flawed. If light can go slightly faster than the speed of light, then Albert Einstein's theory of relativity is flawed. If you would like to disprove Albert Einstein's theory of relativity, then get in a car. Drive faster than 1 mph and shine a flashlight from the rear of the car to the front. The ground speed of that light is faster than the speed of light. Einstein's theory won't stand up to this test. Time won't change. If time did change, the light would go into the past and you would not be able to see it.

Why does speed matter? It's part of the puzzle. Energies have different speeds. It's how the universe works. There is only one universe with one massive amount of energy. Time only ticks by. It doesn't change with speed.

Law 6

Law 6

Energy has different sizes and shapes

Law 6: Opening comments

Energy goes back and forth. Energy goes through matter. Energy is reflected by matter. Energy is changed by matter. Energy moves forward. Energy can move at different speeds. How does energy go through matter? How is energy reflected by matter? How is energy changed by matter? To start to get some answers to these questions, let's think of a pond full of tadpoles. There are big tadpoles and there are small tadpoles. There are fat tadpoles and there are skinny tadpoles.

If you filled the pond with a bunch of bowling balls, most of the tadpoles could swim between the bowling balls. The big fat tadpoles couldn't swim between the gaps. If you filled the pond with BB's, most of the tadpoles could not swim between the BB's. Why can't they swim between the BB's? There is not enough space. What if we used a fishing net with big gaps in the net? Could we catch any tadpoles? Only the big fat ones. The little ones would swim through the gaps. If we used a net with a fine mesh, could we catch any tadpoles? We could probably catch them all. When we go through this law think about the energy as different sizes and shapes of tadpoles. Think about the matter as being the net, or the BB's or bowling balls.

This is a big piece of the jigsaw puzzle. We are grouping pieces of the puzzle. We will be looking at the characteristics of matter and energy. We will be looking at the size and shape of energy.

Light goes through glass

In this observation the energy is light and the matter is a piece of glass. How does light travel through a piece of glass? There are three ways I can see how it could do it. Light either goes through the molecules in the glass or it goes between the molecules of glass or both. Right or wrong? If we go on these three assumptions, we can draw some conclusions. If light could go through the molecules of glass then light should be able to go through any solid matter. We know that is not true. Light goes through very few solids. Therefore, light must go in between the molecules. That would mean that there are gaps between the molecules. These gaps are most likely different from substance to substance. Let's concentrate on glass. The gaps would have a size and shape wouldn't they? Yes, they would. So, for simplicity, let's say the gaps between the molecules in glass are square. So if we have a square hole, what shape would the light need to be? It would need to be square wouldn't it? Yes, it would. So, in our example we are showing light as a square peg and glass with square holes. The light will be able to go through the glass. Energy has different sizes and shapes.

Illustration 6-1: Light goes through glass

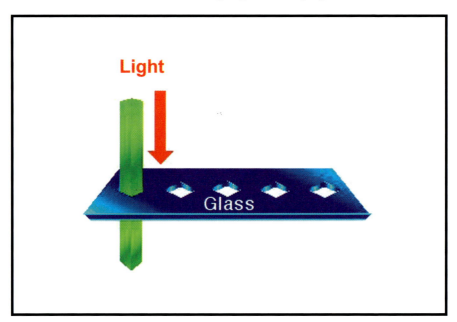

Electricity goes through silver

In this observation the energy is electricity and the matter is a flat piece of silver. How does electricity travel through silver? There are the same three ways as before. Electricity either goes through the molecules in the silver or it goes between the molecules of silver or both. Right or wrong? If we go on these three assumptions, we can draw some conclusions. If electricity could go through the molecules of silver then electricity should be able to go through any solid matter. But we know that is not true. The solids that electricity can't go through are called insulators. Therefore, electricity must go in between the molecules. That would mean that there are gaps between the molecules in silver. These gaps are most likely different from substance to substance. Before when we looked at the glass we said the gaps were like a square hole. Let's say the gaps in silver are round holes. So, if we have a round hole, what shape would the electricity be? It would need to be round wouldn't it? Yes, it would. So, in our example we are showing electricity as a round peg and the silver with round holes. The electricity will be able to go through the silver. Energy has different sizes and shapes.

Illustration 6-2: Electricity goes through silver

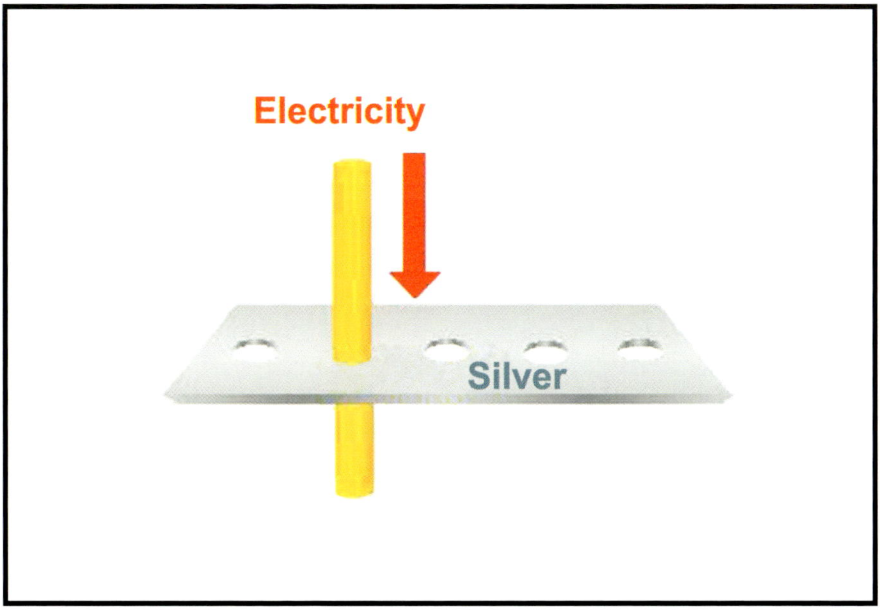

Electricity and light and glass

In this observation the energy is electricity and light. The matter is a flat piece of glass. We are going to use a square peg for light. We will use a round peg for the electricity. The piece of glass will have square holes. If we move the light forward what happens? The light will go through the glass. A square peg will fit in a square hole. Now when we move the electricity forward what will happen? Since electricity is a round peg, it won't go through the glass. A round peg won't fit in a square hole. The electricity will be reflected or changed. Energy has different sizes and shapes.

Illustration 6-3: Electricity and light and glass

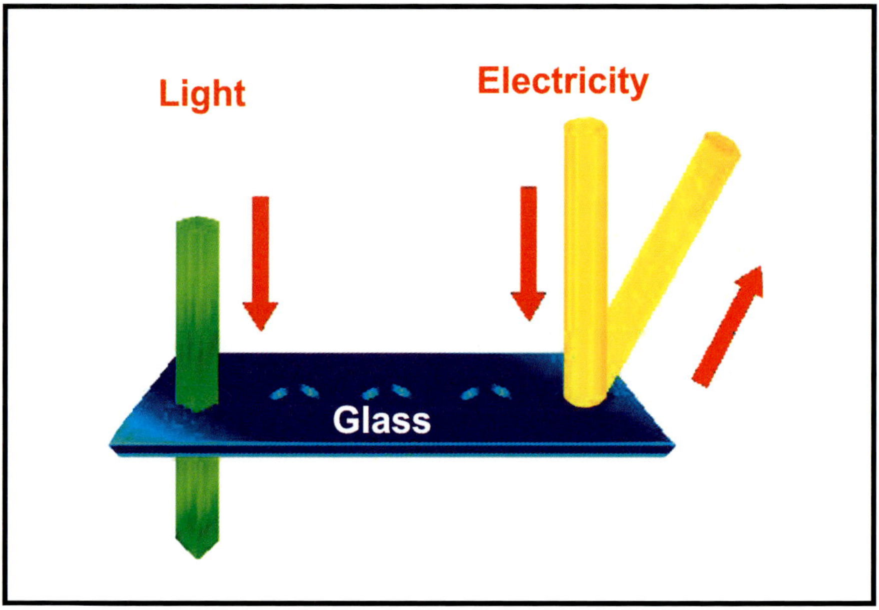

Electricity and light and silver

In this observation the energy is electricity and light. The matter is a flat piece of silver. We are going to use a square peg for light. We will use a round peg for the electricity. The piece of silver will have round holes. If we move the light forward what happens? The light will not go through the silver. A square peg will not fit in a round hole. The light will be reflected or changed. Now when we move the electricity forward what will happen? Since electricity is a round peg, it will go through the silver. A round peg will fit in a round hole. Energy has different sizes and shapes.

Illustration 6-4: Electricity and light and silver

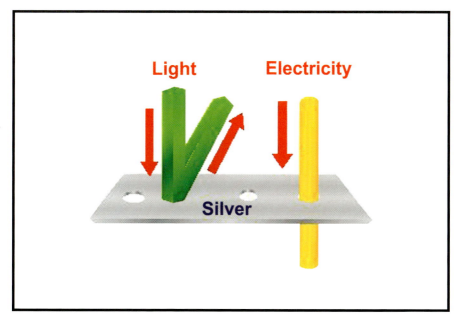

Electricity and light and a mirror

In this observation the energy is electricity and light. The matter is a flat piece of glass with silver on the back. It is a mirror. We are going to use a square peg for light. We will use a round peg for the electricity. The piece of glass will have square holes in the glass and round holes in the silver. If we move the light forward what happens? The light will go through the glass but be reflected or changed by the silver. A square peg will fit in a square hole but not go through a round hole. Now when we move electricity forward what will happen? Since electricity is a round peg, it won't go through the glass. A round peg won't fit in a square hole. The electricity won't even make it to the silver. The electricity will be reflected or changed. Energy has different sizes and shapes.

Illustration 6-5: Electricity and light and a mirror

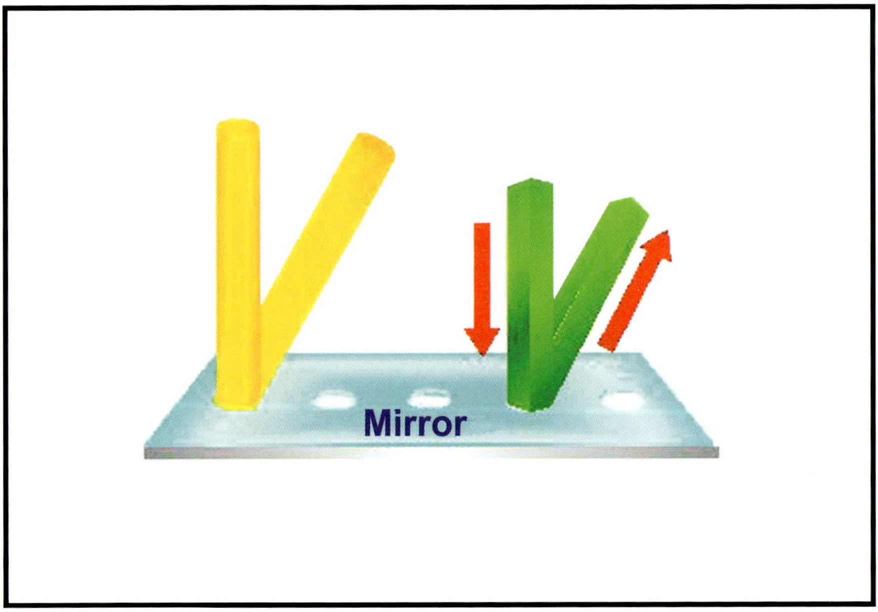

Radio and a mirror

In this observation the energy is a radio wave. The matter is a flat piece of glass with silver on the back. It is a mirror. We are going to use a small round peg for a radio wave. If we move the radio wave forward what happens? The radio wave will go through the glass and through the silver. You can listen to the radio on the other side. What's the conclusion? Radio waves are smaller than electricity or light. So, do the radio waves go between the molecules or go through the molecules? If they go through the molecules of glass and silver then the width of the glass and silver should not make a difference. Radio waves will only travel through a certain thickness of matter. Most likely a radio wave will pass between molecules. Energy has different sizes and shapes.

Illustration 6-6: Radio and a mirror

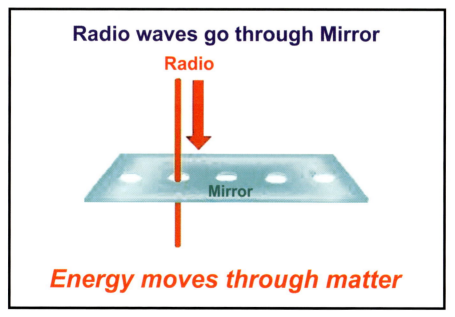

Magnetic energy and a mirror

In this observation the energy is magnetic energy. The matter is a flat piece of glass with silver on the back, the same mirror. We will take a magnet and a piece of iron and put them on each side of the mirror. The magnet and iron will attract each other as if the glass and silver are not even there. Be careful if you do this experiment. If the magnet and iron come together too quickly they will break the glass. The thickness of the glass and silver do not seem to affect how the magnet works. So what is the size and shape of the magnetic energy? They would have to be smaller than light or electricity. Now does magnetic energy go between the molecules or through the molecules or both. Since thickness does not seem to make a difference. Magnetic energy may go between the molecules and through the molecules. Energy has different sizes and shapes.

Illustration 6-7: Magnetic energy and a mirror

Magnetic energy goes through mirror

Magnet

Mirror

Iron

Energy moves through matter

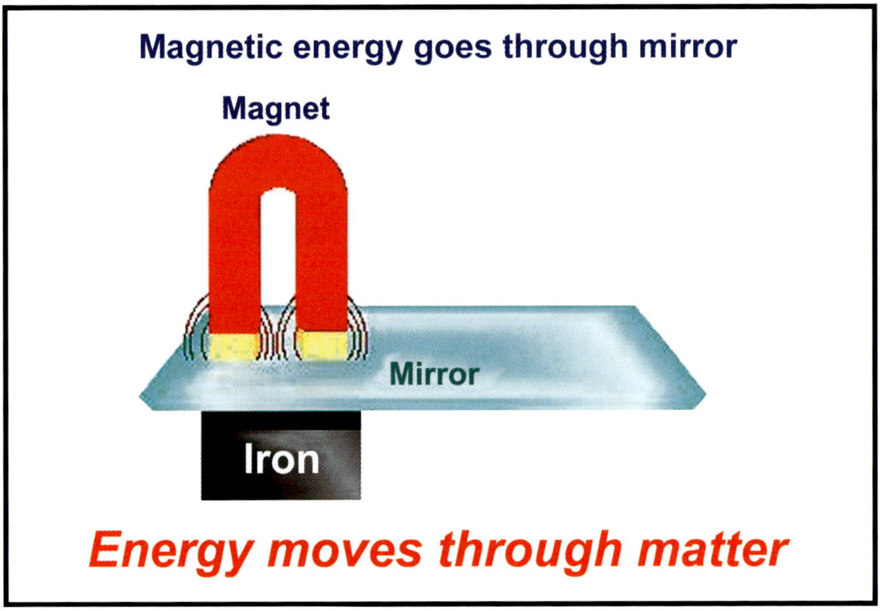

Sound and a mirror

In this observation the energy is sound. The matter is a flat piece of glass with silver on the back, the same mirror. We will have Taddles bark on one side of the mirror. You can hear the sound of Taddles' bark. The bark will be muffled. Sound can travel large distances through solids. Even if the glass is thicker, you may still be able to hear the bark. Sound is changed by the thickness of the glass. What is the size and shape of sound? The size and shape must be smaller than light or electricity because sound will go through both the glass and silver. Does sound go through the molecules or between both molecules? If you change the thickness of the glass, you can still hear the sound. Sound most likely goes between the molecules and through them. Energy has different sizes and shapes.

Illustration 6-8: Sound and a mirror

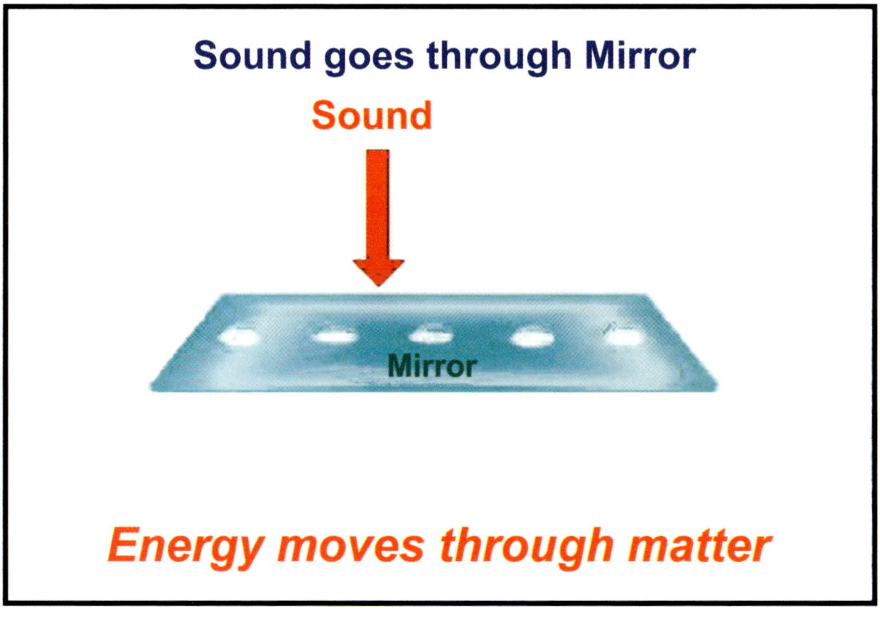

Sound goes through Mirror
Sound

Mirror

Energy moves through matter

Gravity and a mirror

In this observation the energy is gravity. The matter is a flat piece of glass with silver on the back, the same mirror. This time we will suspend Taddles on a scale. We will suspend a mirror above Taddles and lay a mirror under the scale. The gravity will go through the mirror and through Taddles. Taddles will weigh 20 pounds. The gravity will continue on through the second mirror. If you made the mirror above Taddles a mile thick, he would still weigh 20 pounds. If you lay a mile thick mirror under the scale, Taddles would still weigh 20 pounds. Gravity is unaffected by the material or the thickness of matter. Gravity probably goes between the molecules and through the molecules. Gravity may have the thinnest shape and smallest size. Energy has different sizes and shapes.

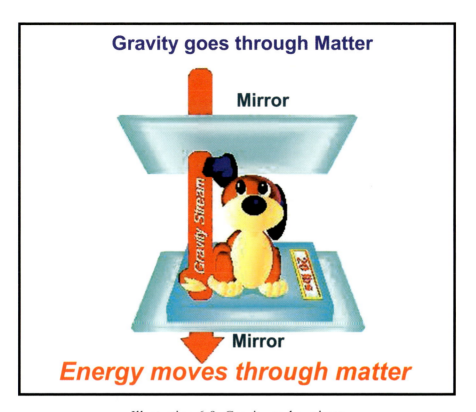

Illustration 6-9: Gravity and a mirror

Several energies and water

In this observation the energy is light, electricity, radio waves, magnetic energy, sound, infrared light and gravity. The matter is pure steam distilled water. Light, radio waves, magnetic energy, sound, infrared light and gravity will travel through this type of water. All of these energies seem to act the same as going through glass. Light and infrared light probably go between the molecules of water. The other energies seem to go between the molecules and through them. Electricity will not go through the pure steam distilled water, there does not seem to be a hole that electricity will fit through. Next we will add a small amount of salt to the water. The light, radio waves, magnetic energy, sound, infrared light and gravity will act as before but, now the electricity will also go through. The salt must change the way the gaps between the molecules are, so electricity will travel through the water and salt mixture. Now add some green food coloring to the water. What happens? Electricity, radio waves, magnetic energy, sound, infrared light and gravity travel through the water, just the same as before. But there is a big difference with the light. Only the green light comes through the water. Why? Even light must have different sizes and shapes. The green food coloring must block all the other colors of light from coming through. Next let's add blue food coloring to the green, salt water. It will almost turn black. Very little light will make it through. Why? Light must have different sizes and shapes. The blue food coloring must block all the green light that the green food coloring didn't block. All the other energies work the same. Try this in a glass of water. Take your remote control and see if it will control the TV through the black water. Infrared must be smaller than any of the other lights. Energy has different sizes and shapes.

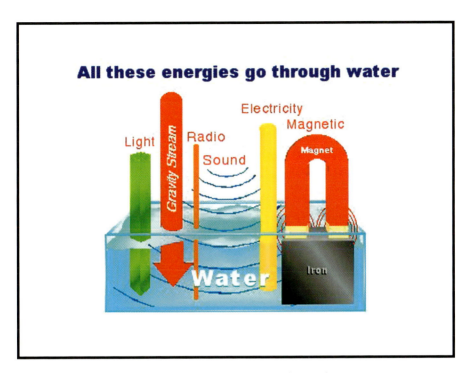

Illustration 6-10: Several energies and water

Electricty, light, glass and silver

In this observation we don't have square pegs or square holes. We don't have round pegs or round holes. We have light which will go through glass. We have electricity which won't go through glass. We have electricity which will go through silver. We have light which will not go through silver.

Light goes through glass but electricity doesn't. Electricity goes through silver but light doesn't. Why? Electricity has a different size and shape than light. Glass has holes that have a different size and shape than silver. Think about it.

Illustration 6-11: Electricty, light, glass and silver

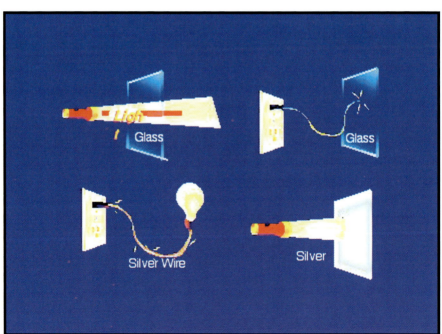

Law 6: Closing comments

I hope this has given you a different way to look at energy. The square pegs and round holes may seem a little far-fetched, but wait until you read further in the book before you come to any conclusions. As we go on, you will see that the energy in the universe is a lot like a pond full of tadpoles and you will see that the matter in the universe is a lot like the net with different sizes of mesh.

Law 7

Law 7

Energy has different states

Law 7: Opening comments

This is one of the biggest pieces of the puzzle. This piece of the puzzle has been completely upside down. We can't see most of the energy in the universe. We haven't put a particle behind each type of energy. In this law, I am going to put a particle behind several different types of energy.

Let's go back to our pond full of tadpoles. There are big tadpoles. There are little tadpoles. There are fat tadpoles. There are thin tadpoles. You can see tadpoles everywhere in the pond. You leave for 2 weeks. You come back to the pond. There are now no tadpoles. There are frogs everywhere. What happened? You know the answer. The tadpoles changed from a tadpole to a frog. The tadpole had a tail and a head. He was stuck in the pond. All this tadpole could do was swim around the pond. After the tadpole changed to a frog he could walk, jump and swim. When the tadpole changed to a frog, he could also leave the pond and go somewhere else. This is a very easy for us to believe because we can see the tadpole change to a frog. Energy is something we can't see. We also can't see it change. Energy changes almost instantly. My ideas will be based on the tadpole, frog, pond approach. If I look and see one type of energy going in an environment and I see another type of energy coming out, my conclusion will be it has changed.

Electricity changes to light

In my first observation the two energies are electricity and light. We will send electricity through a silver wire. Then we will put a light bulb at the end of the wire. The light bulb will give off light. At this point we have two things that supposedly happen. First the electricity is now gone. So where did it go? Secondly, we now have light. Where did it come from? It has commonly been believed that the electricity was a stream of electrons. It has also been believed that light is a stream of photons. Is that how it works, the electrons disappear and photons are created? Put two and two together. The electrons don't disappear. The photons are not created. The electrons are changed into photons! How is this possible? It may be very simple. This would mean that there is at least one particle of energy that can be changed to a different state. So, you can change an electron into a photon. Can you change a photon into an electron? They do with solar cells. Light comes in electricity goes out. Energy has different states.

Illustration 7-1: Electricity changes to light

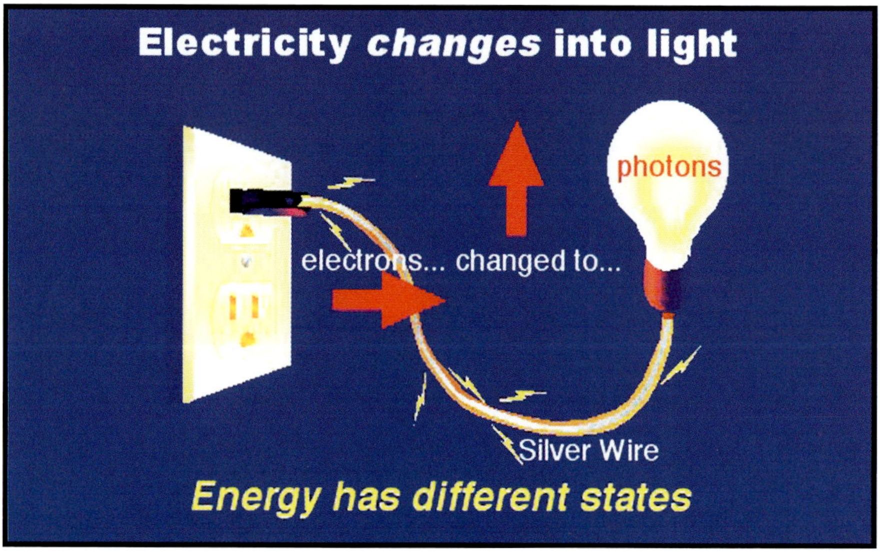

Electricity goes in and light goes out

In my second observation we will imagine what this energy particle may look like and how it may work. We do not have the technology to see the particle. Let's use our brains to think about how a particle looks. I will use my first five laws to base how I think one particle would look. The five laws are "Energy can move through Matter", "Energy can be reflected by Matter", "Energy can be changed by Matter", "Energy moves Forward", and "Energy has Different Sizes and Shapes". Based on these things I imagine the particle to look like a tadpole. It has a head and it has a tail. The head has a shape that will determine how it will act. The tail will determine how the energy particle will move forward. It may be that simple. I will use this tadpole approach in all my examples. I will name this particle the "Tadtron". What would the tail and the head on electricity look like? Lets say the head is round and the tail is shaped so it can snake its way in between molecules. This tadtron is now in the electricity state. The tadtron is snaking its way through the silver wire and it comes to the light bulb. The light bulb changes the tadtron's state from the electricity state to the light state. How is this done? It's the way it works. Now the tadtron has a new shape of its head and a new shape for its tail. Its head may now be square. Its tail now has a shape to fly straight ahead. What's the conclusion? Electricity goes in and light goes out. Energy has different states.

Illustration 7-2: Electricity goes in and light goes out

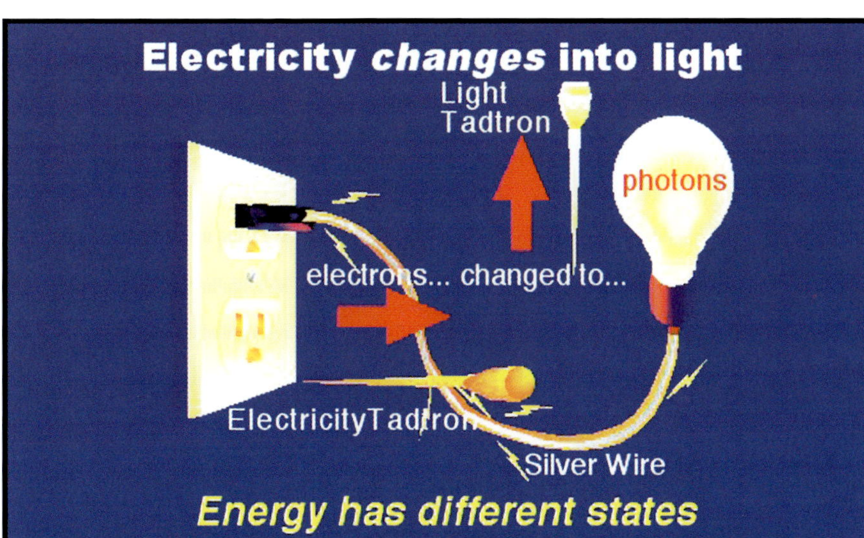

Electricity changes to radio

In my third observation the 2 energies are electricity and a radio wave. In this example a tadtron will move through a silver wire as electricity and go to a radio transmitter. The radio transmitter changes the state of the tadtron from electricity to a radio wave. The head is most likely thinner than light or electricity. The tail is now told to move in a wave pattern. This wave pattern can be controlled by the transmitter. What's the conclusion? Electricity goes in and radio waves come out. Energy has different states.

Illustration 7-3: Electricity changes to radio

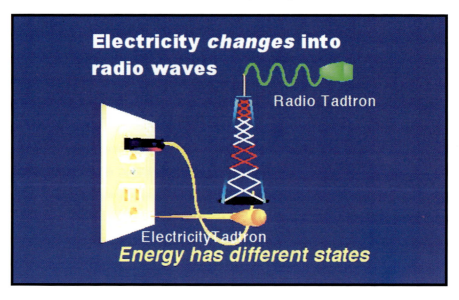

Magnetic energy changes to electricity

In my fourth observation the 2 energies are a magnetic field and electricity. When you pass a magnet by a coil of wire, a current of electricity is produced. This process is called a generator. Supposedly, the electricity comes from electrons out of the magnetic field. These were identified as being of the same type of energy particles. They were called electrons. Why aren't you shocked by a magnet the same as you would be from a current of electricity in a wire? The magnetic energy is in a different state. The shape of the head of the tadtron in the magnetic state is probably very thin. The shape of the tail is for flying in a circular pattern. Some of the tails may have a different shape to fly in a larger circular pattern. When you pass the magnet along the metal, the magnetic tadtrons change to electricity tadtrons. Energy has different states.

Illustration 7-4: Magnetic energy changes to electricity

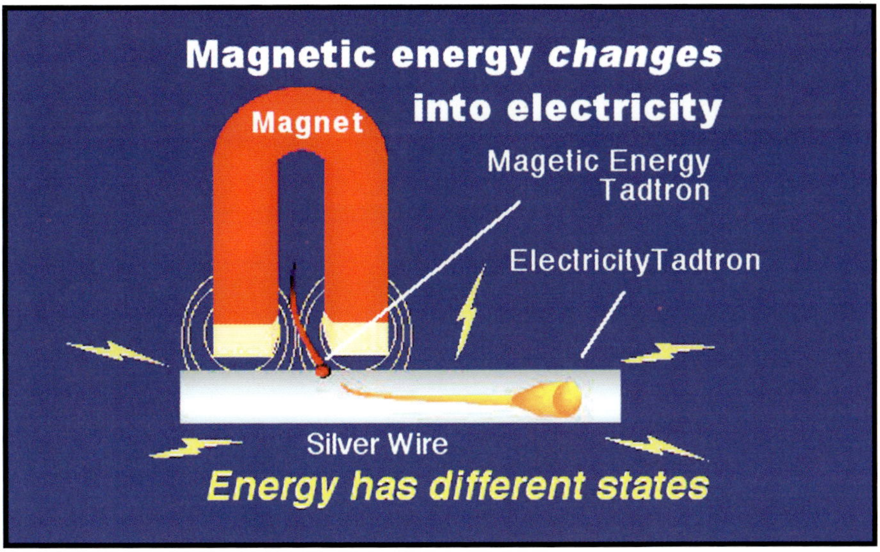

Magnetic energy changes to electricity and then light

    In my fifth observation I will use 3 states of energy. Our energies will be magnetic energy, electricity, and light. We will run a generator, hook it to a wire, then hook the wire to a light bulb and then we will shine the light on a flat black surface. When we pass the magnet along the wire, a magnetic tadtron changes states to an electricity tadtron. The electricity tadtron travels down the wire to the light bulb. The electricity tadtron is changed to the light tadtron state. The light tadtron will travel to the flat black surface where 97 % of the light is now "absorbed" in the black surface. It's as simple as a tadpole changing to a frog. But, we have some big holes to fill. Magnets in generators never run out. Where do the replacement magnetic tadtrons come from? The logical conclusion is that they come from an energy state that we have not identified. Where did the light tadtron go? The logical conclusion is that it was converted to another energy state that we have not identified. Energy has different states.

Illustration 7-5:
Magnetic energy changes to electricity and then light

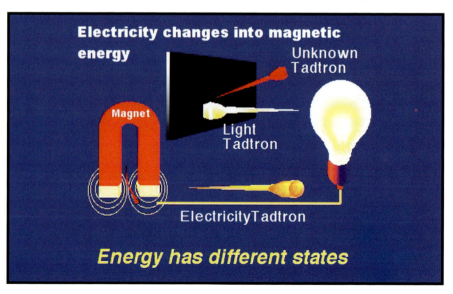

Electricity changes into magnetic energy

In this observation I will use electricity and an electromagnet. In the illustration assume it is direct current coming out of the outlet. An electricity tadtron goes in the wire and goes in a circle around the iron nail. This makes an electromagnet. The electricity tadtron state changes to the magnetic tadtron state. The iron nail is now a magnet. In the previous observation, we changed a magnetic tadtron to an electricity tadtron. Now we changed an electricity tadtron to a magnetic tadtron. The electricity tadtron could have traveled thousands of miles down the wire to get there. That tadtron could have come from another magnet or a battery. Energy has different states.

Illustration 7-6: Electricity changes into magnetic energy

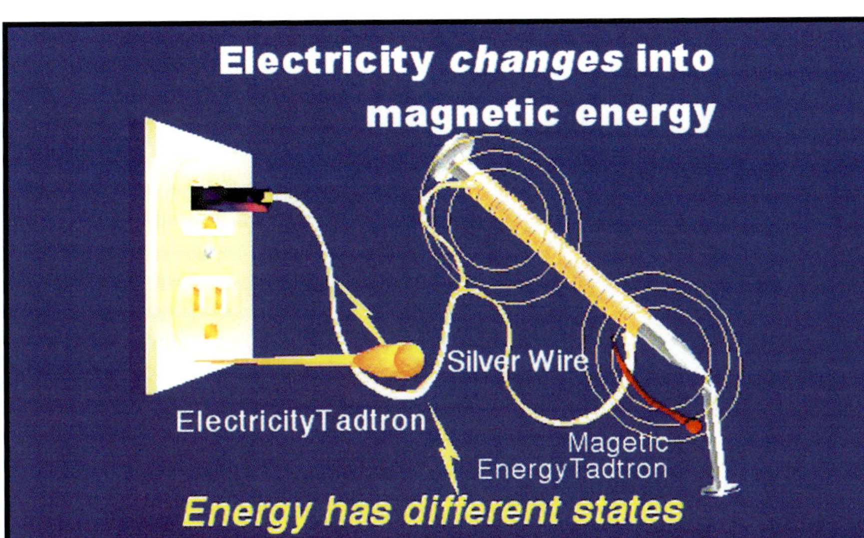

Law 7: Closing comments

We started this law by thinking about a pond full of tadpoles. We talked about how the tadpole was stuck in the pond. We talked about how the tadpole could change into a frog. Then we took that concept and applied it to energy. We put a particle behind several types of energy. We named that particle the tadtron. We showed how a single tadtron particle could change from one state to the next. This is where we start to leave the tadpole behind. The tadpole can only swim in the water. The tadpole can only change into a frog. The tadtron can change into thousands of different energy states. The tadtron can change back to a state it already was. A magnetic tadtron can change to an electricity tadtron and then change back to a magnetic tadtron. A frog cannot change back to a tadpole. Tadtrons can swim through solids, liquids, gases and through space.

We can now think of the universe as a pond full of tadtrons. There is a single particle that changes state, behind every type of energy. Once you turn this piece of the puzzle up, the rest of the puzzle gets much easier to put together.

Law 8

Law 8

Atoms are energy particles traveling in circles

Law 8: Opening comments

You can't put a jigsaw puzzle together with one of the pieces upside down. It was taught in grade school science classes that an atom was made of protons, neutrons and electrons. It was taught in high school science classes that an atom was made of protons, neutrons and electrons. It was taught in college science classes that an atom was made of protons, neutrons and electrons. No one has ever shown me a picture of a proton. No one has ever shown me a picture of a neutron. In electrons, we have seen the effects of electricity.

The vast majority of all our science is based on the proton, the neutron, and the electron idea. What if the proton doesn't exist? What if the neutron doesn't exist? What if this piece of the puzzle is upside down? Could this be the big stumbling block in physics?

In this chapter I am going to show you that the proton does not exist. I am going to show you that this piece of the puzzle is upside down. I am going to show you a much better model of an atom. Physics and chemistry will soon make a lot more sense to you. This is a huge piece of the puzzle.

Ye old atom

The subatomic theory lays things out pretty much as follows. You have a proton that has a positive charge. Then you have an electron which has a negative charge. The positive charged proton attracts the negative charged electron. The electron rotates around the proton in an orbit like the moon orbits the earth. If you look at this logically, then why doesn't the electron stick to the proton? When you have an element that has two protons and two electrons, why don't the two positively charged protons blow apart? Don't two positive charges push each other apart? Should the two electrons push each other apart? Then you have the neutron. It is a neutral charge. What holds it in place? This model was molded after the Earth and moon. Has anyone ever seen a picture of a proton? We have only seen drawings and illustrations. I am going to refer to this as the old theory.

I am going to give you a new model of a molecule that makes a lot more sense. Before I do that I want to show the old model of a hydrogen atom. The next drawing is an illustration of a hydrogen atom with the old theory. There is a small circle with a "P" in the center. This represents the proton. There is a small circle with an "E". This represents the electron. The large circle with the arrows is the orbital path of the electron. The proton theoretically has a positive charge. The electron theoretically has a negative charge. The positive and negative charges theoretically attract each other. In quantum mechanics the attraction is called weak force. The electron theoretically orbits the proton. The weak force theoretically holds the electron in this orbit. The electron's orbit is now a shell around the proton. There is now a lot of space in the atom. This model was obviously molded after the earth and the moon. Gravity holds the moon in place.

There have been pictures taken of molecules with different types of microscopes. The fuzzy outside shell is all that has ever been filmed. No one has ever been able to take a picture of a proton inside of a molecule.

An idea that has been around for a long time is collapsed matter. When the weak force doesn't work any longer or there aren't any electrons to go around the proton, then you have collapsed matter. It's a proton without an electron shell. Collapsed matter is theoretically very dense. The pull of gravity will become very strong. Theoretically, nothing can escape this gravity pull.

This dense matter is supposed to be a black hole. The gravity is supposed to be so strong that light can't escape it. The black hole is a theory from another theory. Soon you will see, with this new model of an atom, that you can't have collapsed matter.

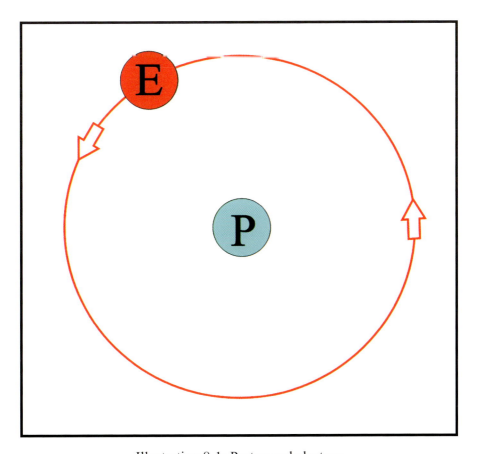

Illustration 8-1: Proton and electron

Key ring, Key and 2 circling Tadrons

So what is the building block of all matter? Atoms are the building blocks. What does an atom look like? First of all atoms have mass or a weight. With sophisticated equipment scientists have weighed atoms. Scientists have weighed a single electron as well. The electron weighs more than a thousand times less than an atom. Let's work with the electron weighing 1/1000 of an atom for simplicity. If an electron is a tadtron in the electron state, then a tadtron weighs 1/1000 of an atom. Everything is made up of energy. Then, by doing some simple math, that would mean there are 1000 tadtrons in a single atom. The tadtrons would have to be in a stable atomic state. All 1000 of the tadtrons are held together in a very stable way. So, how would they all be held together? Let's go back to our fourth law, which is"energy moves forward". Let's think of the tadtron as a dog. If the dog acted like light he would take off and run away from you. You would not be able to make an atom out of a tadtron traveling in a straight line. If the tadtron traveled in a wave pattern, the results would be the same as a straight line. However, if the dog ran around in a circle he would go nowhere. Think of this as a dog chasing his tail. If energy moves forward, what movement keeps things in the same position? The answer is a circle. Magnets have energy moving in a circle. What if these circles are much smaller? That would mean we would have 1000 tadtrons traveling in small circles. Something would have to hold them together. What? The answer may be in your pocket. Pull out your key ring and look at it. You have a circle holding a circle. The key ring is a circle. Each key on the key ring has a circle. How often do your keys fall off? It's a pretty stable configuration to keep track of your keys. I believe the key ring is the configuration of all atoms.

In the next illustration I have a key ring with a key on it. The key ring is the gray circle. There is one key on the key ring. The key is yellow. You can put a lot of keys on the key ring. Next to the key ring is the start of my new model of an atom. I have a tadtron traveling in a tight circle in black. The tadtron is long and thin. The tail overlaps itself. This is the key ring. This will hold the whole atom together. Next is a red tadtron traveling in a circle. The red tadtron acts like the key on the key ring. It is a much larger circle than the key ring tadtron.

We will now go on and add more keys to the key ring.

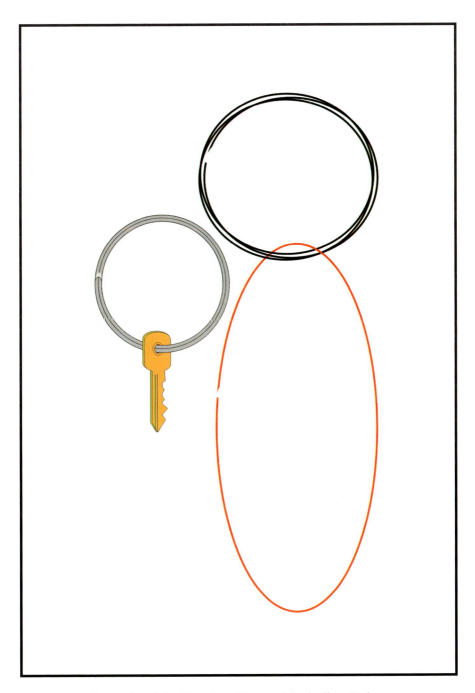

Illustration 8-2: Key ring, Key and 2 circling Tadrons

The key ring atom

The key ring atom has a single tadtron traveling in a very small circle at its center. Its tail and its head are connected and overlap like a key ring. It is a "ring". I will name this state of energy as a "Proton Ring". The proton ring is the center of all matter. The hole in the center is what all gravity streams will go through. Each proton ring will have a series of other tadtrons circling through the proton ring. These tadtron rings I will name "Electron Rings". The electron rings will circle through the proton ring. There is a limited number of electron rings. They will tightly align themselves next to each other until the Proton ring is full. This is a single atom. It is the building block of all matter. How many Electron rings are there? Most likely there are over a thousand electron rings. But let's go with the 1000 we said earlier. That would mean the proton ring would weigh 1/1000th of an atom. There would be 999 electron rings, which would weigh 999/1000th of an atom. We will have a total of 1000 tadtrons in the atom, giving it its atomic weight. For simplicity in our illustration we are only going to show 60 electron rings around the proton ring. They are multicolored for visual effects.

On the next page is the key ring atom. This configuration would be hydrogen. At the center of the atom is a pink proton ring. This holds the atom together just like a key ring holds your keys. The electron rings form all the way around. What is the shape of this atom? It's the shape of a donut. It is also the shape of an inner tube that goes in a tire. The electron rings are shown as a complete circle. We will show them this way in most of the rest of our illustrations. It is much easier to draw and explain this way. What is the size of the electron rings? I will give that answer later when I explain hot and cold.

I am going to do some reverse engineering. Historically, one of the best ways to understand how something works is to take it apart and look at all the pieces. In the next illustration we are going to show what happens to an atom when it is broken apart. How will we do that? We will do a hydrogen bomb simulation. The next several pages will be 4 steps of the key ring atom being broken apart.

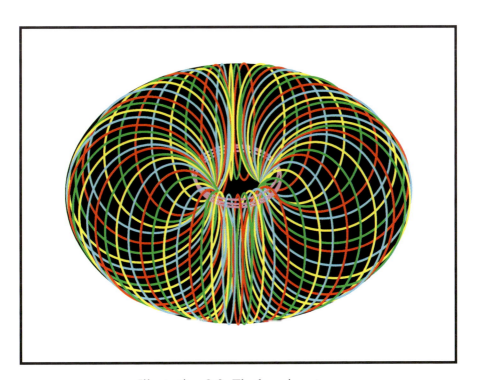

Illustration 8-3: The key ring atom

Trigger of a hydrogen explosion

We have the technology to produce a hydrogen bomb. When a hydrogen bomb is exploded, the first thing that happens is the proton ring is broken. The proton ring is changed to another state. What state? Well, what comes out of a hydrogen bomb when it explodes? The answers are light, radio waves, electromagnetic waves, radiation, and a huge sound wave. The proton ring in this next example is changing to light as it exits the atom. As the proton ring changes to light, the rest of the electron rings now start to break apart. In the bomb itself this would be the trigger of the explosion.

Think of your key ring with the keys on it. Someone is taking the key ring and straightening it out. As the key ring gets straighter the keys start falling off.

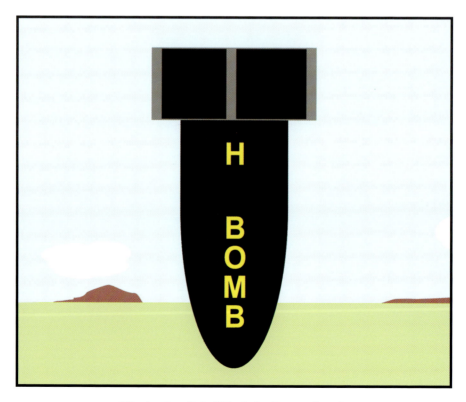

Illustration 8-4: Whole hydrogen bomb

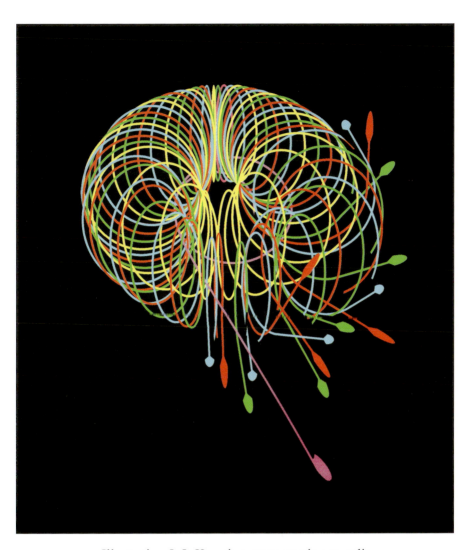

Illustration 8-5: Key ring atom starting to split

Energy starts coming out of the hydrogen bomb

In these illustrations the bomb starts to blow apart. This is a fraction of a second after the bomb is triggered. The electron ring has changed to light. Nothing holds the electron rings in place. The electron rings are now changing states to light, radio, radiation and sound. This is the start of the hydrogen explosion. The energies that travel at the speed of light are leaving the center of the atom. The remaining electron rings at the center are slower and will travel out after the speed of light energies.

Think of your key ring with the keys on it. Someone is taking the key ring and straightening it out. At this point all the keys are falling off.

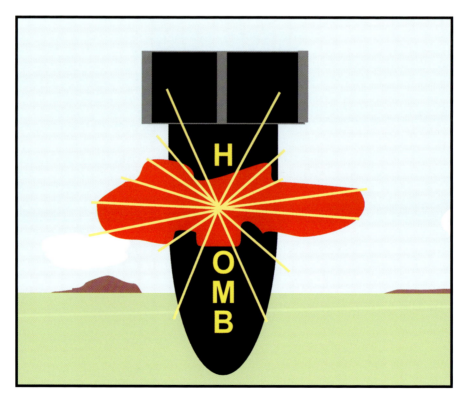

Illustration 8-6: Hydrogen breaking apart

Illustration 8-7: Proton ring almost gone

Massive flash of light

In these illustrations the bomb has exploded. If we were watching, we would see a massive flash of light. The last of the electrons rings are breaking apart. The electron rings have almost all changed to light, radio waves, electromagnetic waves, radiation, and sound waves. At this point the hydrogen bomb has exploded.

Think of your key ring again. The key ring is gone. The keys are loose and on their own. Nothing holds them together.

Illustration 8-8: Flash of light

Illustration 8-9: Last of proton rings breaking apart

Sound comes out of hydrogen bomb

In these illustrations we will show the last of the electrons rings changing to sound. The yellow corkscrews are the sound. This is the most destructive part of the bomb. The sound waves now branch out at the speed of sound. For a few miles out they destroy everything in their path. You would be able to see a large mushroom cloud. Why is sound a corkscrew? We will cover that in another law.

Think of your key ring one last time. The key ring is gone. All the keys are loose and free to go wherever they want.

Illustration 8-10: Mushroom cloud

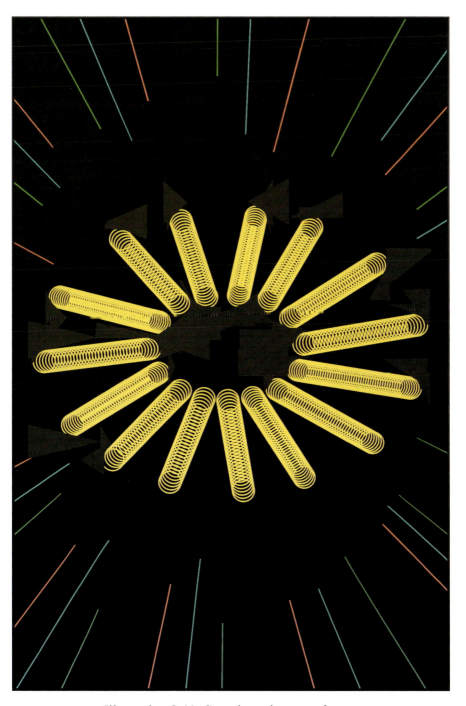

Illustration 8-11: Sound coming out of atom

Law 8: Closing comments

When you break the model of an atom apart, you can track every single particle as it changes state.

Now think about the old theory. What's in a proton? How does the proton turn into all that energy? Does the proton create all that energy when it is split apart? There where two things that accounted for this release of energy. Those two things were nuclear fission and nuclear fusion. Nuclear fission is the dividing of an atom. That is what we have done with our hydrogen bomb example, we divided an atom. Nuclear fusion, in theory, is the combining of protons and neutrons. Fusion is supposed to produce more energy than fission. Fusion has never made sense to me. I am not going to try and explain it.

In a hydrogen bomb, isotopes of hydrogen are used. The isotopes are deuterium and tritium. In the next chapter we will show deuterium and tritium. We will see how these are actually divided and are not fused. I will discuss fusion in the next chapter.

A nuclear bomb is usually made of an isotope of uranium. It is called uranium 235. Nuclear fission, in a nuclear bomb, is the dividing of uranium 235. Part of the uranium protons or neutrons are broken into smaller pieces. We will discuss this more in the next chapter.

With the key ring atom you can't have collapsed matter. If you can't have collapsed matter, then you can't have a black hole. Black holes were thought to be a part of the puzzle of how the universe worked. It is an imaginary piece. Collapsed matter doesn't exist. Black holes don't exist.

What did we change? In the old theory there was 1 proton and 1 electron. The bulk of the weight or the mass was in the proton. In this theory there is 1 proton ring and many electron rings. The bulk of the weight or the mass is in the electron rings. This piece of the puzzle was upside down. I have now turned it right side up. The puzzle can now be put together.

The key ring atom is central to the rest of this theory. It is a huge piece of the puzzle. The key ring atom doesn't operate like the sun and the moon.

A single particle holds about one thousand particles in place. There is a hole in the center of the key ring. This hole will answer a lot of questions on gravity. The key ring atom is the building block of all matter. In the next chapter we'll start building the fundamental pieces of our universe.

Law 9

Law 9

An element is a chain of proton rings

Law 9: Opening comments

The elements are one of the biggest parts of the puzzle. In this chapter we are going to build 20 of the first known elements. We are going to put them together with the key ring atom. Once we do this, we will have the frame work, or outside edges, of our jigsaw puzzle. After we build this frame, the rest of the jigsaw puzzle of the universe will be much easier to put together and understand.

In this chapter I am going to discuss the old theory of the atoms and elements. Second, I am going to lay out my new atoms and new elements. Third, I am going to show a metal molecule. Fourth, I am going to show a couple of items that should be called elements. Then I am going to discuss nuclear fusion and nuclear fission. This is the basic construction of our known universe. This is one of my favorite chapters in the book.

I don't believe a neutron exists. I do believe that a neutron particle exists. At the end of this law I will explain this, at the same time I explain my concepts behind nuclear fission.

Old oxygen molecule

No one has ever seen an atom. No one really knows what an atom looks like. The old subatomic theory has 3 things in an atom. They have protons, neutrons, and electrons. Protons and neutrons are at the core of the atom. The electrons rotate around the protons and neutrons like the moon orbits around the earth. Protons are supposed to have positive charges. Neutrons are supposed to have neutral charges. Electrons are supposed to have negative charges. The positive charge in the protons is supposed to hold the negative charged electrons in orbit. In quantum mechanics this is known as weak force. Another force holds the protons and neutrons together. It is known as strong force in quantum mechanics. Strong force has never made sense to me. If you have 2 positive charges, they should repel each other. The positive charged protons at the center of each element should blow apart, in the old subatomic theory. Elements were given an atomic number based on how many protons are in the atom. Elements were also given atomic weights. The atomic weight is the weight of all protons plus the weight of all neutrons. Anything that does not have the same number of protons and neutrons is an isotope. There are many isotopes of each element.

There are different levels of orbits for the electrons. These orbits are called shells. Electrons in the outer orbit can be shared with other elements to produce a bond. The illustration on the next page is a model of oxygen with the old theory. It has 8 protons and 8 neutrons. It has 8 electrons. Six of the electrons are in the outer orbit. Two of these 6 electrons can be shared with 2 other elements to make compounds.

This model is the basic construction behind all of the elements. There are many problems with this model. Sharing electrons to make bonds doesn't make sense. Heat and cold is caused by the protons and neutrons vibrating. There are many isotopes. That means there are many elements that don't have an equal number of protons and neutrons.

Illustration 9-1: Old oxygen molecule

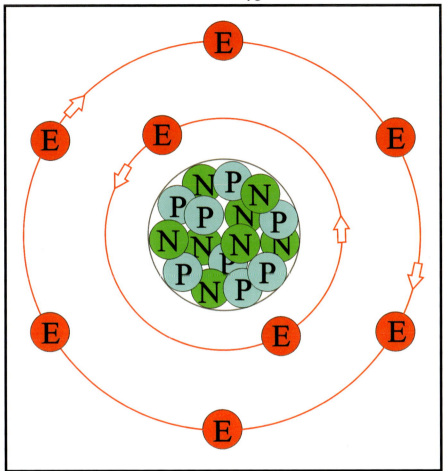

More on the old theory

Many have tried since the dawn of time to identify all the materials that are on this earth. We know rocks are different than water. We know gold is different than silver. Materials can be separated. Materials can be combined. The materials act differently when they are separated. Materials act differently when they are combined.

More on the old theory

Scientists have tried to separate materials until they could not be separated any further. Once a material was separated, it would be identified as an element. Each of the elements were given names. Some of the elements were gases. Some of the elements were liquids. Some of the elements were solids. Depending on what the temperature is, makes a difference on whether an element is a solid, a liquid or a gas.

After identifying many elements, scientists tried to classify the elements. They tried to put them in an orderly way. This was done in the late 19th century and early 20th century. They came up with what is known as the periodic table. I remember seeing the periodic table on the wall in my science class when I was in high school. It was treated as fact.

The periodic table was based on the idea that the number of protons determined an element's atomic number. Each element also has an atomic weight. Elements also have neutrons. The atomic weight of an atom was based on the number of protons and neutrons. In the previous illustration of oxygen we had 8 protons and 8 neutrons. The atomic number of oxygen would be 8. The atomic weight of oxygen would be 16. There were supposed to be the same number of neutrons as protons. There were over a hundred elements that were identified.

A century has passed since the periodic table was first developed. Many new elements have been discovered that are not on the current periodic table. To account for this, scientists came up with something called isotopes. Isotopes are elements that have a different number of neutrons than the standard on the periodic table. For example, you could find oxygen with an atomic weight of 17. It would then have 9 neutrons instead of 8. The neutron is the neutral item that makes up for what you can't explain. Isotopes and neutrons are big red flags. When you see red flags, something may be wrong at the core of the old theory. Let's question the old theory.

Let's ask a couple of simple questions. Has anyone ever proven that a proton exists? Has anyone ever proven that a neutron exists? I have never seen any proof that either exists. Any pictures of any atoms, that I have seen, only show the outside of the atom. It is always a shadow or a bump. You can't get to the inside to see the proton or the neutron. In law 8 it shows that the proton ring holds an atom together. Obviously, I don't believe a proton exists. I don't believe a neutron exists either. So, to classify the elements, I have to give you a new way of looking at elements.

We used a key ring as a simple example to explain an atom. So, how do we make elements? When your key ring is full, the only way to add more keys is to chain multiple key rings together. It's not complicated. Take some key rings and put them together. You can come up with many different configurations. You can make chains, spirals, circles, and legs. I started putting key rings together to see what I could come up with.

I decided to take the first 20 elements that are commonly known and make them into multiple key rings. I played with different configurations. To make a configuration I would start with the atomic weight from the periodic table. Then I would see how that element acted chemically. Some really cool patterns started to emerge. I noticed there were a number of key rings at the center of each element. There were some key rings that formed shells around the centers. There were some legs that came out of the center or off of the shells.

Elements bond with other element, which is known as a chemical bond. There is a wide variety in the way elements bond. Some elements will not bond with any other element. Some elements can bond with up to 4 other elements. Some elements will bond with one element and not another. Valence is the bonding characteristics of an element. If an element has a valence of 0 it will bond with nothing. If an element has a valence of a -1 it will bond with another element that has a positive valence. If an element has a valence of +1 it will bond with an element that has a negative valence. If an element has a valence of +-4 it will bond with up to 4 other elements that have positive or negative valences.

For my construction, if any element has a positive valence, I put 2 proton rings in each leg. If it has a negative valence, I put 3 or more proton rings in each leg. Then I took the atomic weight and rounded the number to the closest integer. I used that integer for the total number of proton rings. Then I constructed the elements and came up with a molecular number. I named this the molenum. C stands for center. S stands for shell. L stands for legs. You will see the molenums as we describe the elements. I will show 5 elements at a time and then discuss them in the next several pages. There is a 2D and 3D image of each of the elements.

Elements 1 through 5

HYDROGEN
Element 1 is Hydrogen. It has an atomic number of 1. Hydrogen has an atomic weight of 1. Hydrogen has a valence of +1. The 2D image is a single circle. It has a molenum of C1.

HELIUM
Element 2 is Helium. It has an atomic number of 2. It has an atomic weight of 4. It has a valence of 0. It doesn't bond with any other elements. The 2D image is a circle of 4 proton rings. The molenum is C4.

LITHIUM
Element 3 is Lithium. It has an atomic number of 3. It has an atomic weight of 6.9. It has a valence of +1. It will bond with 1 other elements that has a - valence. The 2D image is a circle of 5 proton rings and a single leg with 2 circles. The molenum is C5-1L2. The C5 is for 5 proton rings in the center. The 1L2 is for 1 leg with 2 proton rings.

BERYLLIUM
Element 4 is Beryllium. It has an atomic number 4 and an atomic weight of 9. It has a valence of +2. It will bond with 2 other elements that have a - valence. The 2D image is a circle of 5 proton rings and 2 legs with 2 circles each. The molenum is C5-2L2. The C5 is for 5 proton rings in the center. The 2L2 is for 2 legs with 2 proton rings in each leg.

BORON
Element 5 is Boron. It has an atomic number of 5. It has an atomic weight of 10.8. It has a valence of +3. It will bond with 3 other elements that have a - valence. The 2D image is a circle of 5 proton rings and 3 legs with 2 circles in each. The molenum is C5-3L2. The C5 is for 5 proton rings in the center. The 3L2 is for 3 legs with 2 proton rings in each leg.

Atomic Number	2 D Image	Element	Valence	Atomic Weight	Molecule Numbering
1		Hydrogen	+1	1	C1
2	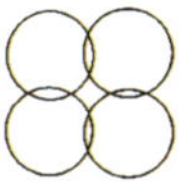	Helium	0	4	C4
3	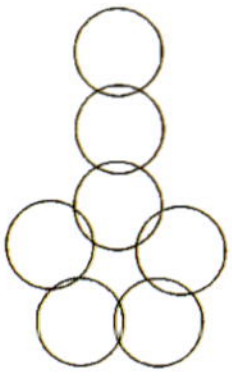	Lithium	+1	6.9	C5-1L2
4	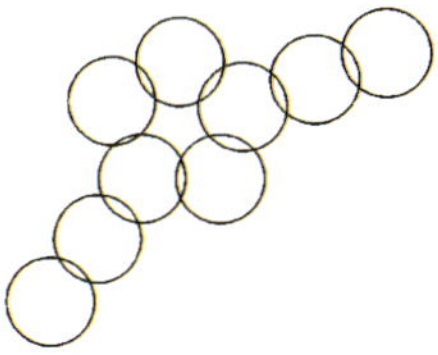	Beryllium	+2	9	C5-2L2
5	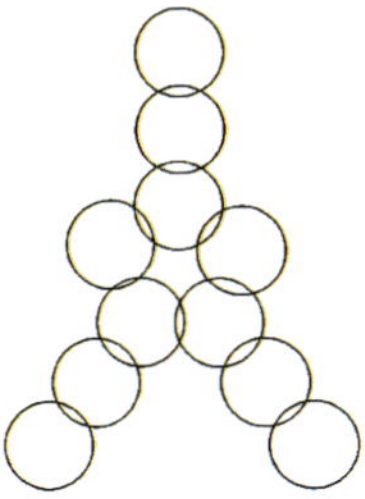	Boron	+3	10.8	C5-3L2

Illustration 9-2: Elements 1 through 5

These next illustrations show the first 5 elements in 3 dimension. For simplicity all the proton rings will be a made from a heavier line than the electron rings. We will only use 4 electron rings on each proton ring. If we used 1,000 it would be too cluttered and you could not see the proton ring. These 4 electron rings will let you see the inner workings of each element.

HYDROGEN

This is a 3D image of Hydrogen. Hydrogen is normally a gas. The melting point of Hydrogen is -259 C. The boiling point is -252 C. By weight, Hydrogen is the ninth most abundant element on the earth. It makes up a little less than 1% of the earth. Hydrogen is the lightest of all the elements. Hydrogen is present in most forms of life, in coal, in petroleum, and in all acids. We used 60 electron rings on Hydrogen, rather than 4 as on all the rest. You will see how busy things would look if we used 60 on all elements.

Illustration 9-3: Hydrogen

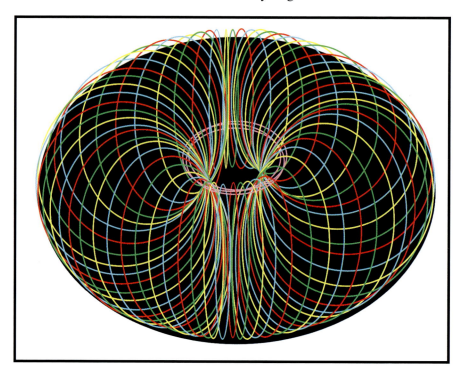

HELIUM

This is a 3D image of Helium. Helium is a gas that is odorless and tasteless. Helium is the next lightest substance below Hydrogen. Notice how round Helium looks. This will come into play when we talk about solids, liquids, and gasses. Helium's boiling point is -268C. Now we need to talk about density. Density is the degree of compactness of a substance or the mass per volume. When we add more proton rings to an element, the density of that element may increase. The increase is due to the overlapping of electron rings. The more overlap of electron rings in an element, the higher the density will be. Helium has a few overlapping electron rings. Helium will not freeze. It has a condition that is called super fluidity.

Illustration 9-4: Helium

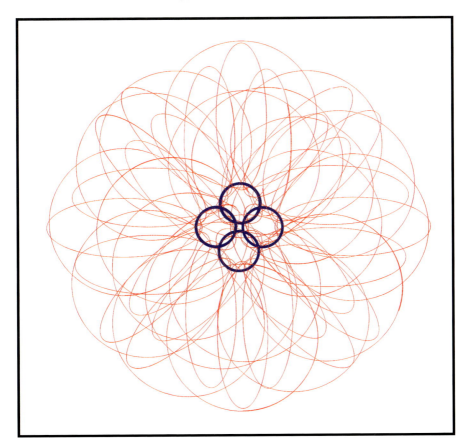

LITHIUM

This is a 3D image of Lithium. Lithium is considered to be a metallic element. It is normally white. It is the lightest of all the solids. The melting point is 186 C. The boiling point is 1317 C. Notice the density increase in Lithium over Helium. Adding the additional proton ring in the center and adding the 1 leg causes more electron rings to overlap. This would account for why Lithium is a solid. The leg is a connection point for a bond with another element.

Illustration 9-5: Lithium

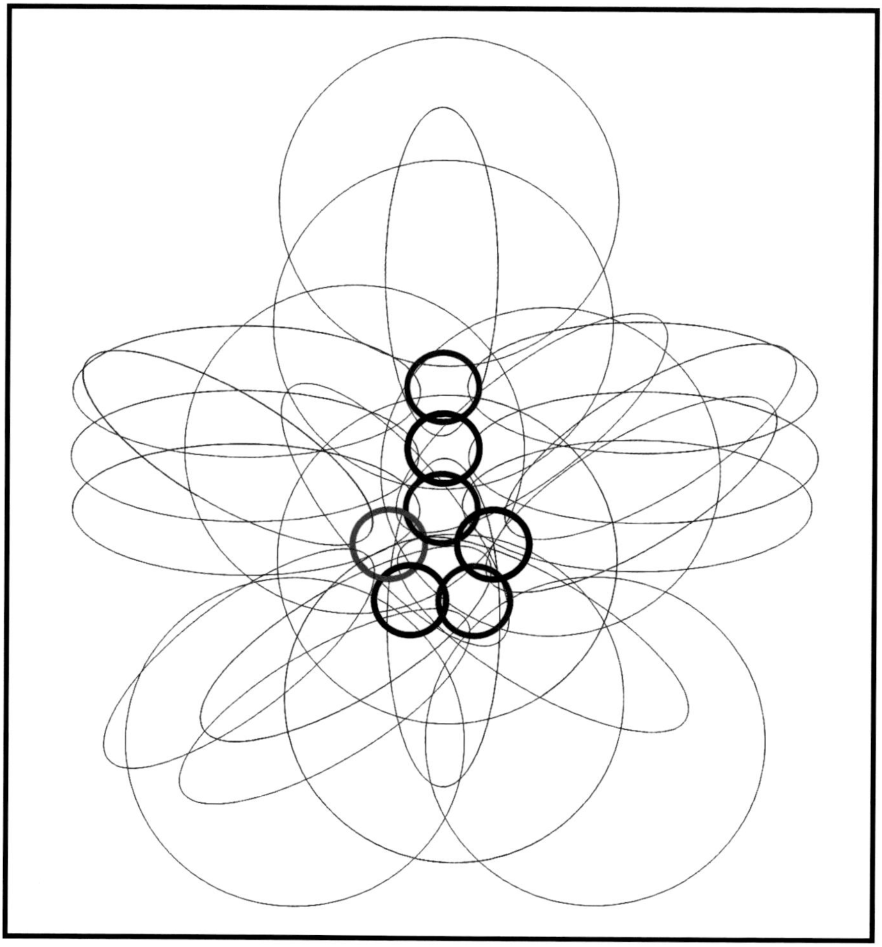

BERYLLIUM

This is a 3D image of Beryllium. Beryllium is considered to be a light metallic element. The melting point is 1285 C. The boiling point is 2507 C. Notice the density increase in Beryllium over Lithium. Adding the additional leg causes more electron rings to overlap. There are 2 legs for 2 connection points for a bond with 2 other elements.

Illustration 9-6: Beryllium

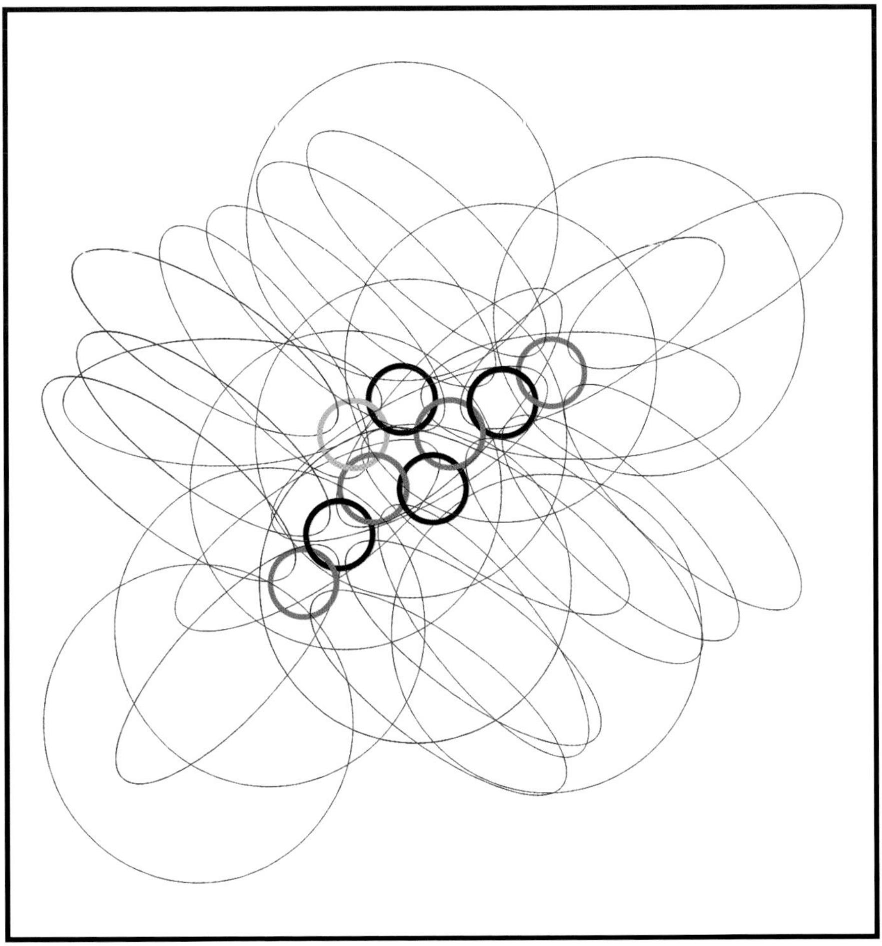

BORON

This is a 3D image of Boron. Boron is known as a semi metallic element. It has been known in nature as Borax. The melting point and boiling point are above 2200 C. Density increases as we add the additional leg to Beryllium. There are 3 legs that allow for 3 connection points for 3 bonds to 3 other elements.

Illustration 9-7: Boron

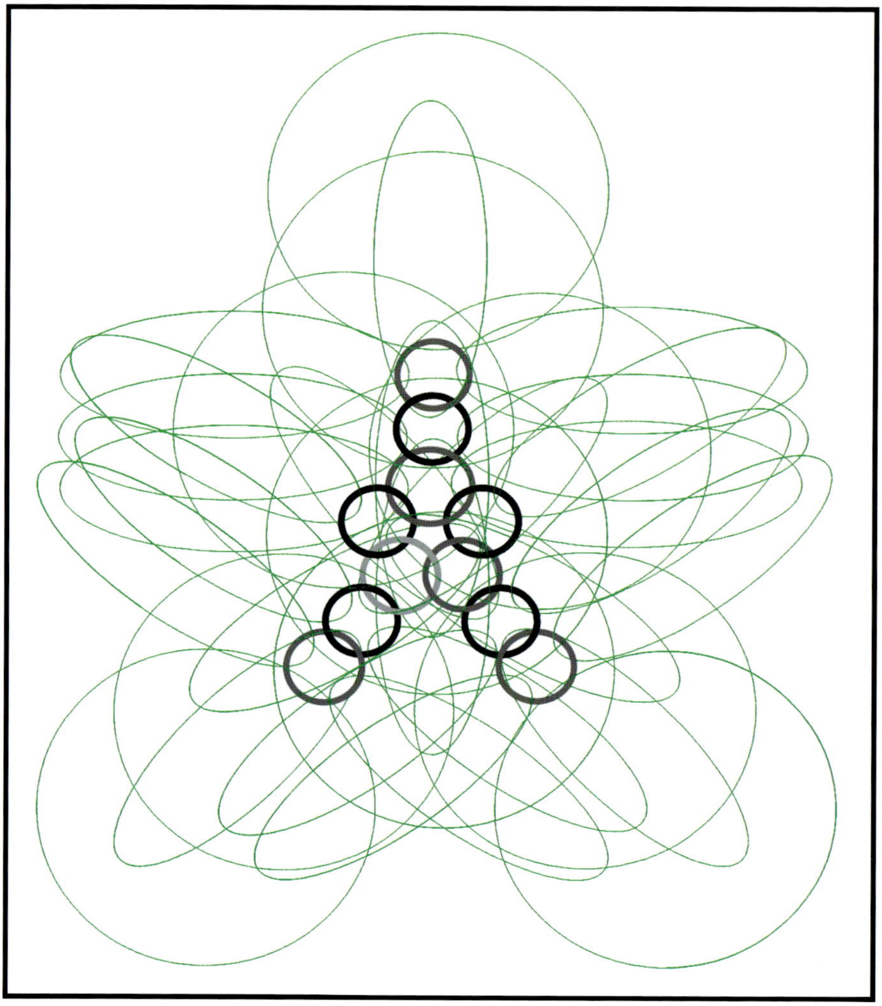

Recap of molecules 1 through 5

Why do elements 3, 4, and 5 only have 2 proton rings in each leg? I came up with this to go along with a positive valence. Because they have a positive valence, they will not easily combine with another element that has a positive valence. For example Lithium has a +1 valence. Boron has a +3 valence. They will not easily combine to form a bond. Elements with 2 proton rings in a leg do not like to bond with other legs that have 2 proton rings in them. Why don't they like to combine? It looks like there may be a collision with the overlapping electron rings from the proton rings in the center. We will do bonding in law 10. This is part of the foundation of chemical bonding.

Elements 3, 4, and 5 have a center with 5 proton rings. Why? It works! The numbers come out.

What about isotopes? An isotope is something that doesn't weigh the same as the regular element. For example what if some Beryllium was found that had an atomic weight of 10 instead of 9. How would you account for this? Easy, add a proton ring in the center or add a proton ring onto one of the legs. Depending on where the proton ring is, will cause the Beryllium isotope to act different chemically. It will bond differently. It will weigh differently.

Hydrogen, theoretically, has 2 isotopes. They are Deuterium and Tritium. In the old theory Deuterium has an extra neutron. In the old theory Tritium has 2 extra neutrons. In my theory Deuterium would have a molenum of L2. In my theory Tritium would have a molenum of L3. I would classify them as elements and not isotopes. We will talk about these after element 20.

Let's move to elements 6 through 10. We will now see a negative valence.

Elements 6 through 10

CARBON

Element 6 is Carbon. It has an atomic number of 6. It has an atomic weight of 12. It has a valence of +-4. It will bond with 4 other elements that have a + or - valence. The 2D image is a circle of 4 proton rings and 4 legs with 2 circles each. The molenum is C4-4L2. C4 is for 4 proton rings in the center. The 4L2 is for 2 legs with 2 proton rings in each leg.

NITROGEN

Element 7 is Nitrogen. It has an atomic number of 7. It has an atomic weight of 14. It has a valence of -3. It will bond with 3 other elements that have a + or - valence. The 2D image is a circle of 5 proton rings and 3 legs with 3 circles in each. The molenum is C5-3L3. The C5 is for 5 proton rings in the center. The 3L3 is for 3 legs with 3 proton rings in each leg.

OXYGEN

Element 8 is Oxygen. It has an atomic number of 8. It has an atomic weight of 15.9. It has a valence of -2. It will bond with 2 other elements that have a + or - valence. The 2D image is 16 circles in a chain. The molenum is 1L16. The 1L16 is 1 leg with 16 proton rings

FLUORINE

Element 9 is Fluorine. It has an atomic number of 9. It has an atomic weight of 18.9. It has a valence of -1. It will bond with 1 other element with a + or - valence. The 2D image is a circle of 12 proton rings and a single leg with 7 proton rings. The molenum is C12-1L7. The C12 is for 12 proton rings in the center. The 1L7 is for 1 leg with 7 proton rings.

NEON

Element 10 is Neon. It has an atomic number of 10. It has an atomic weight of 20. It has a valence of 0. It will not bond with other elements. The 2D image is a circle of 20 proton rings. The molenum is C20. The C20 is for 20 proton rings in the center.

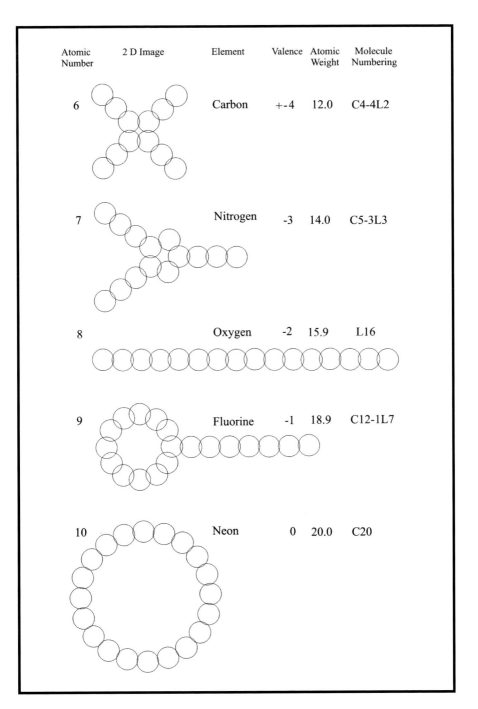

Illustration 9-8: Elements 6 through 10

The next 5 illustrations show elements 6 through 10 in 3D. All proton rings are heavier lines than the electron rings. There will be 4 electron rings on each proton ring.

CARBON

This is a 3D image of Carbon. Carbon is considered to be a nonmetallic solid. Carbon may have over 1 million different bond configurations with different elements. Carbons natural state may be a diamond or graphite. Notice the way the legs on Carbon chain together. The legs are almost 3 long instead of 2. This may account for why Carbon has a + or - valence.

Illustration 9-9: Carbon

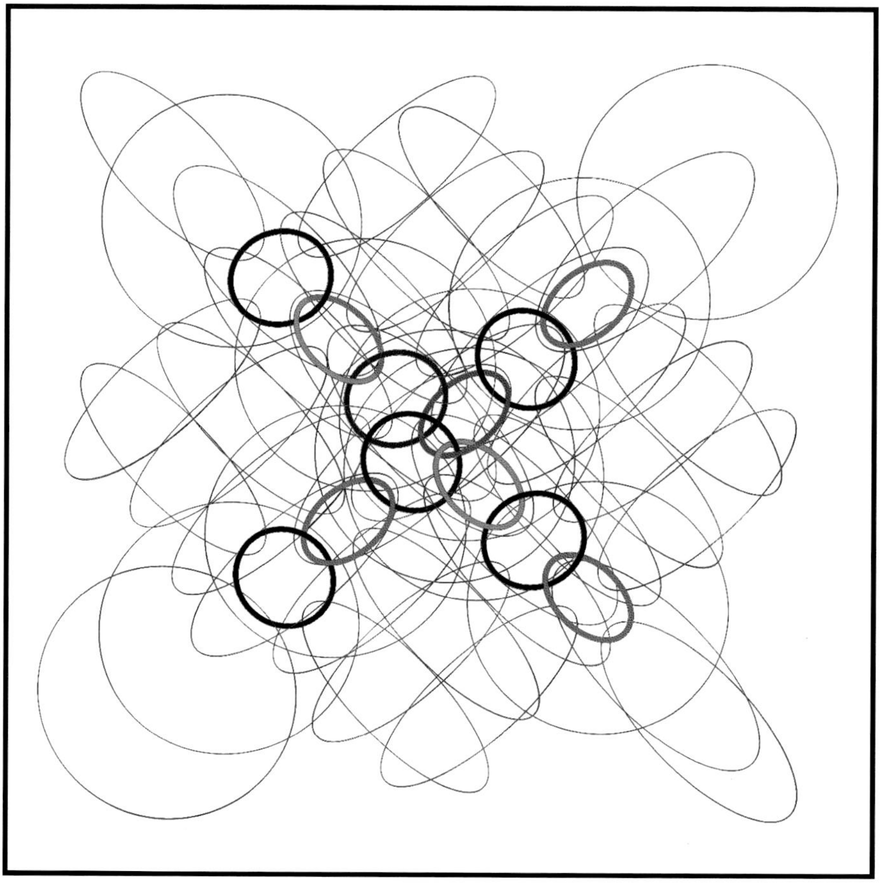

NITROGEN

This is a 3D image of Nitrogen. Nitrogen is a gas that is odorless, tasteless and colorless. It makes up about 78% of our atmosphere. If you compare Nitrogen to Boron, you will only see 1 difference. The legs have 3 proton rings in each leg rather than 2. As the legs get farther from the center proton rings there will be fewer electron rings that overlap. This will lower the density. This should explain why Nitrogen is a gas. Nitrogen makes up many compounds. It is widely used in explosives.

Illustration 9-10: Nitrogen

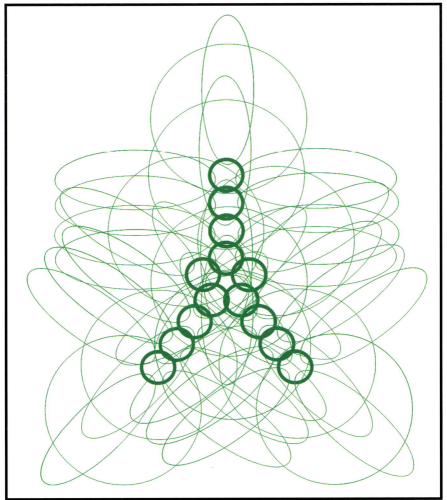

OXYGEN

This is a 3D image of Oxygen. Oxygen is a gas that makes up about 21% of our atmosphere and makes up about half of the earth's crust in various compounds. Oxygen is the most abundant element. If you look at the 3D image, you will notice that Oxygen is a chain of 16 proton rings. Oxygen will be able to bend and bond to almost any other element. Oxygen boils at -182 C and freezes at -218 C.

Illustration 9-11: Oxygen

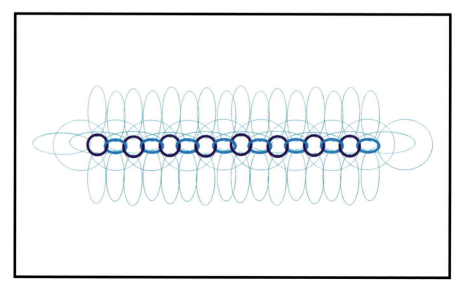

FLUORINE

This is a 3D image of Fluorine. Fluorine is a greenish yellow gas that is very reactive to other elements. Fluorine boils at -188 C and freezes at -223 C. The 3D configuration of Fluorine could be shaped as a dish. This could account for it being a gas. There could be other molenums that could work. Possible configurations could also be a C16-1L3 or C12-1L7.

Illustration 9-12: Fluorine

NEON

This is a 3D image of Neon. Neon is an inert gas. It does not form bonds with any other elements. Neon boils at -246 C and freezes at -248 C. Notice the nice round circle it forms. Very few electron rings overlap which would account for why it is a gas. There are no legs for it to bond with.

Illustration 9-13: Neon

Recap of molecules 6 through 10

This is where you could most likely get the most excited about this new theory. You can look at the molecules and see how they would react in the environment. Look at Carbon, it has 4 places it can connect to other elements. Look at Nitrogen, it has 3 connection points. Look at Neon, it has no connection points. Look at Oxygen, it has 2 connection points. Oxygen is like a log chain, it is able to bend to many different shapes. The legs are longer in some of these elements. That makes them act and look different.

What about isotopes? One of the most well known isotopes is Carbon 14. The Carbon shown has 12 proton rings. To make Carbon 14 just add 2 proton rings. You could add them on 2 of the legs. You could make the center bigger and add more proton rings to it. For isotopes of Oxygen just add or subtract proton rings off of either end.

What about any other elements? Could there be other elements? Absolutely, yes! There could easily be elements between Neon and Helium. There may be several inert gases that are a circle of proton rings like Helium and Neon. These elements would have a number of proton rings between 5 and 19. There may be some elements that are like Oxygen but with fewer proton rings. They would just be a shorter chain.

Let's move to the next 5 elements.

Elements 11 through 15

SODIUM

Element 11 is Sodium. It has an atomic number of 11. It has an atomic weight of 22.9. It has a valence of +1. It will bond with 1 other element that has a - valence. The 2D image is a single circle in the center, 8 circles around the center, 12 circles around the 8 and 1 leg with 2 circles. The molenum is C1-S8-S12-1L2. The C1 is for 1 proton ring in the center. The S8 is for a shell of 8 proton rings. The S12 is for the second shell of 12 proton rings. 1L2 is for 1 leg with 2 proton rings.

MAGNESIUM

Element 12 is Magnesium. It has an atomic number of 12. It has an atomic weight of 24.3. It has a valence of +2. It will bond with 2 other elements that have a - valence. The 2D image is a circle of 5 proton rings, 15 circles around the 5, and 2 legs with 2 circles in each leg. The molenum is C5-S15-2L2.

ALUMINUM

Element 13 is Aluminum. It has an atomic number of 13. It has an atomic weight of 26.9. It has a valence of +3. It will bond with 3 other elements that have a - valence. The 2D image is a circle of 6 proton rings, 15 circles around the 6, and 3 legs with 2 circles in each leg. The molenum is C6-S15-3L2.

SILICON

Element 14 is Silicon. It has an atomic number 14. It has an atomic weight of 28. It has a valence of +-4. It will bond with 4 other elements that have a + or - valence. The 2D image is a circle of 4 proton rings and 4 legs with 3 circles each . The molenum is C4-S12-4L3. C4 is for 4 proton rings in the center. S12 is for the shell of 12 proton rings. 4L3 is for 4 legs with 3 proton rings in each leg.

PHOSPHORUS

Element 15 is Phosphorus. It has an atomic number of 15. It has an atomic weight of 30.9. It has a valence of -3. It will bond with 3 other elements that have a + or - valence. The 2D image is a circle, surrounded by 6 circles, surrounded by 15 circles and 3 legs with 3 circles in each. The molenum is C1-S6-S15-3L3. The C1 is for 1 proton ring in the center. S6 is for the first shell of 6 proton rings. S15 is for the second shell of 15 proton rings. The 3L3 is for 3 legs with 3 proton rings in each leg.

Atomic Number	2 D Image	Element	Valence	Atomic Weight	Molecule Numbering
11	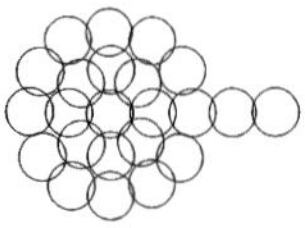	Sodium	+1	22.9	C1-S8-S12-1L2
12	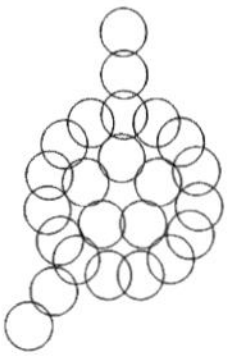	Magnesium	+2	24.3	C5-S15-2L2
13	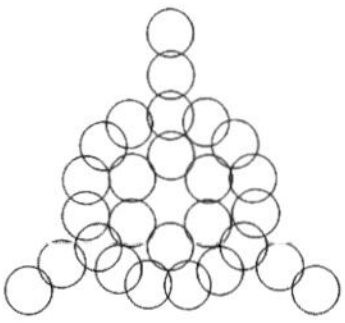	Aluminum	+3	26.9	C6-S15-3L2
14	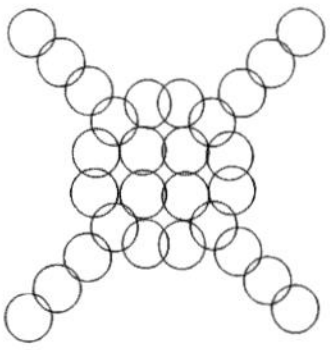	Silicon	+-4	28.0	C4-S12-4L3
15	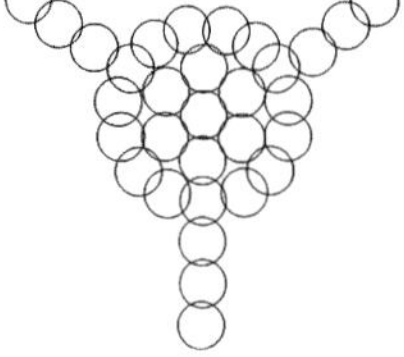	Phosphorus	-3	30.9	C1-S6-S15-3L3

Illustration 9-14: Elements 11 through 15

SODIUM

This is a 3D image of Sodium. Sodium is a soft silvery, white metal. It can be cut with a knife. It is mainly found as table salt in nature. Sodium melts at 97 C and boils at 883 C. The earth's crust consists of about 2.8% Sodium. This makes Sodium the sixth most abundant of the elements. Notice the overlapping electron rings. This is why Sodium is a solid at room temperature.

Illustration 9-15: Sodium

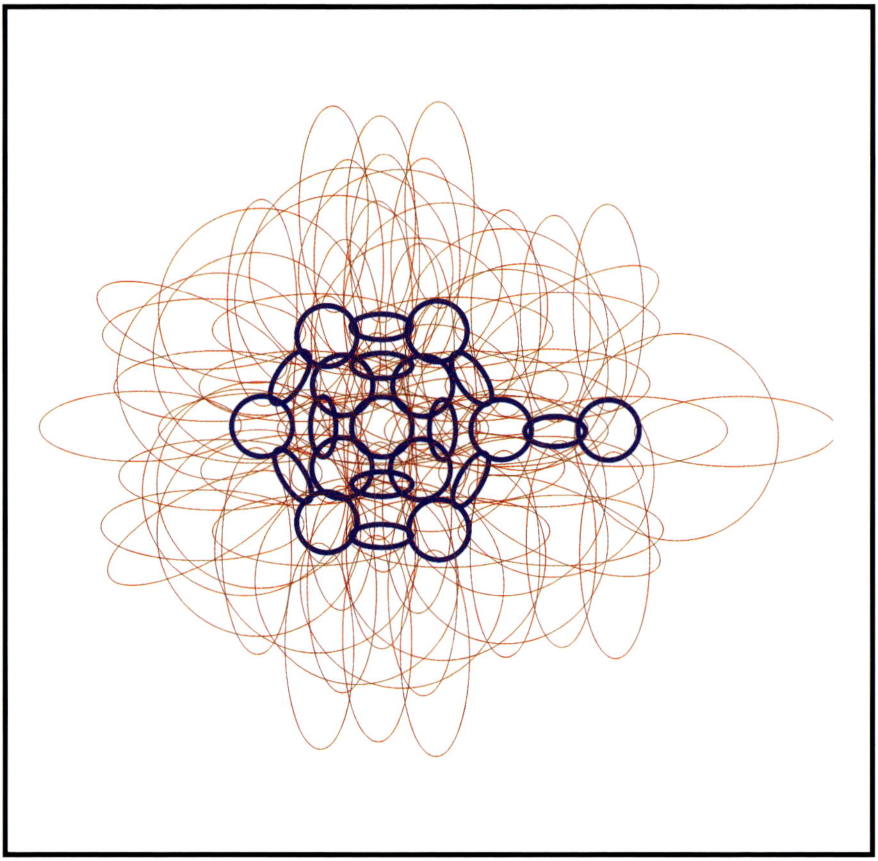

MAGNESIUM

This is a 3D image of Magnesium. Magnesium is a silver white metal. It is the lightest of the structural metals. It is alloyed with other metals to make it stronger and to tarnish less. Magnesium melts at 650 C and boils at 1107 C. Magnesium burns with a very bright flash of light. Magnesium is a solid at room temperature. It has several overlapping electron rings.

Illustration 9-16: Magnesium

ALUMINUM

This is a 3D image of Aluminum. Aluminum is a light silver-colored metal. It is not found naturally in its metal form. It is the third most abundant element in the earth's crust. The earth's crust is about 8% Aluminum. Aluminum melts at 660 C and boils at 2477 C. Aluminum has more overlapping rings than Magnesium. This would account for why it is heavier than Magnesium.

Illustration 9-17: Aluminum

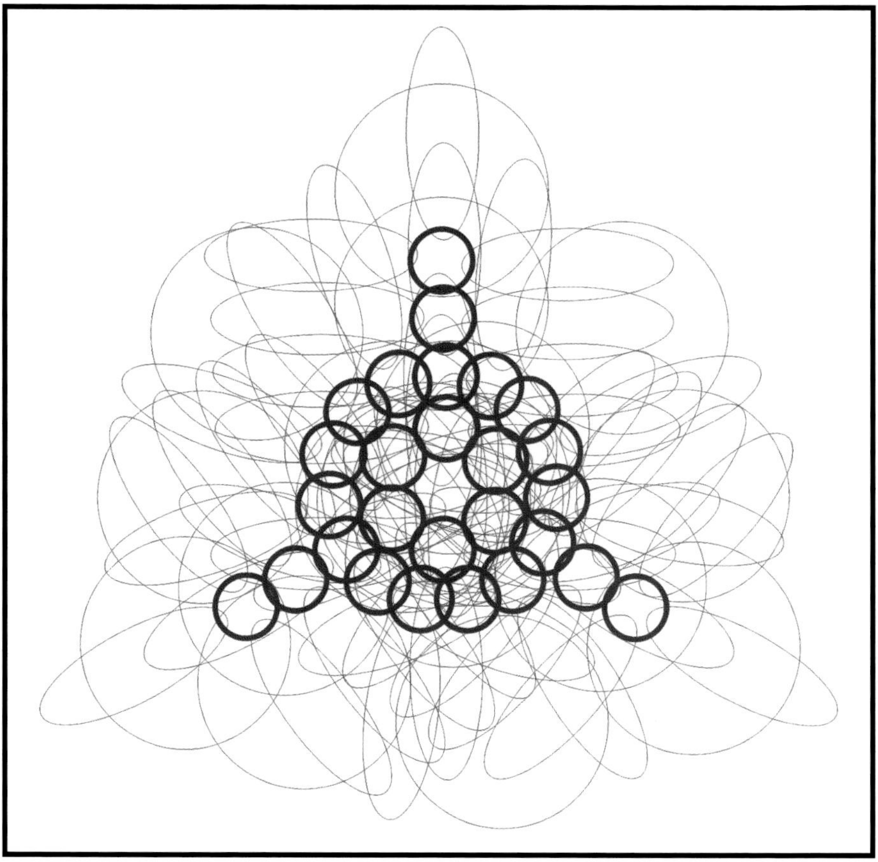

SILICON

This is a 3D image of Silicon. Silicon is like Carbon. It bonds with many other elements. Silicon is the second most abundant element. It is not found in its natural form in nature. Silicon has a boiling point of 2600 C and a melting point of 1420 C. There are many electron rings that overlap in Silicon. This makes it a solid at room temperature.

Illustration 9-18: Silicon

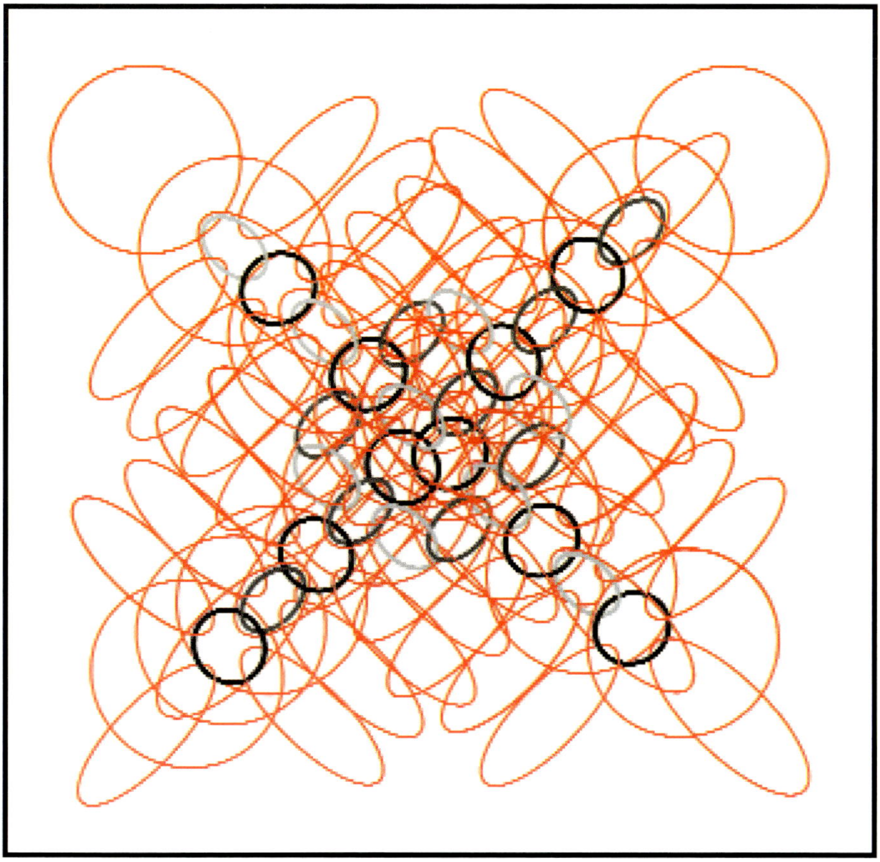

PHOSPHORUS

This is a 3D image of Phosphorus. Phosphorus is a little odd in the forms it takes. It can be a white or red in its ordinary form. Phosphorus boils at 380 C and melts at 44 C. You can see by the overlapping rings it is a solid at room temperature. Phosphorus is used to make matches and fertilizer.

Illustration 9-19: Phosphorus

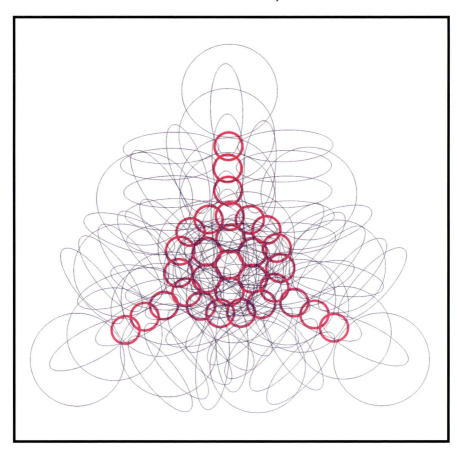

Recap of molecules 11 through 15

We have moved to elements with more proton rings. It becomes more difficult to understand how the centers of the molecules are chained together. It was fun working with several different configurations.

Isotopes are easier to understand. What will happen when you add a proton ring to the center or onto the leg? The element will act and weigh slightly different. As long as the configuration has about the same number of proton rings in the center and the same number of legs, it is going to be classified as the same element.

One of the things to do to arrange these elemental configurations is to match the centers to their density. More dense elements have more proton rings at their center. The number of proton rings corresponds to the atomic weight. I matched the number and length of the legs to the valence. If you line up these configurations with the old periodic table, you will notice the vertical alignment is due to elements having the same number and length of legs. Compare Silicon and Carbon. Both have 4 legs. The main difference is Silicon has more proton rings in its center. Silicon weighs more than Carbon and acts differently.

The next 5 elements are going to make a little departure from what we have seen so far. Think of building a house. You have a shell on the outside. You have rooms on the inside. If you build more rooms inside, you will use more materials. The next elements are like houses that are the same size outside, yet they have a different number of rooms inside.

Elements 16 through 20

SULFUR

Element 16 is Sulfur. It has an atomic number 16. It has an atomic weight of 32. It has a valence of -2. It will bond with 2 other elements that have a + or - valence. The 2D image is 5 circles in the center surrounded by 15 circle and 2 legs of 6 circles. The molenum is C5-S15-2L6. C5 is for 5 proton rings in the center. S15 is for a shell of 15 proton rings. 2L6 is for 2 legs with 6 proton rings in each leg.

CHLORINE

Element 17 is Chlorine. It has an atomic number of 17. It has an atomic weight of 35.4. It has a valence of -1. It will bond with 1 other element with a + or - valence. The 2D image is a circle of 12 proton rings, a second set of 16 proton rings, and a single leg with 7 circles. The molenum is C12-S16-1L3. The C12 is for 12 proton rings in the center. The S16 is for a shell of 16 proton rings. The 1L7 is for 1 leg with 7 proton rings.

ARGON

Element 18 is Argon. It has an atomic number 18. It has an atomic weight of 39.9. It has a valence of 0. It will not bond with other elements. The 2D image is a circle of 40 proton rings. The molenum is C40. The C40 is for 40 proton rings in the center.

POTASSIUM

Element 19 is Potassium. It has an atomic number of 19. It has an atomic weight of 39. It has a valence of +1. It will bond with 1 other element that has a - valence. The 2D image is a single circle in the center, 8 circles around the center, 12 circles around the 8, 16 circles around the 12 and 1 leg with 2 circles. The molenum is C1-S8-S12-S16-1L2. The C1 is for 1 proton ring in the center. The S8 is for a shell of 8 proton rings. The S12 is for the second shell of 12 proton rings. The S16 is for the third shell of 16 proton rings. 1L2 is for 1 leg with 2 proton rings.

CALCIUM

Element 20 is Calcium. It has an atomic number of 20. It has an atomic weight of 40. It has a valence of +2. It will bond with 2 other element that have a - valence. The 2D image is a circle of 8 proton rings, 12 circles around the 8, 16 circles around the 12, and 2 legs with 2 circles in each leg. The molenum is C8-S12-S16-2L2. The C8 is for 8 proton rings in the center. The S12 is for the first shell of 12 proton rings. The S16 is for the second shell of 16 proton rings. 2L2 is for 2 legs with 2 proton rings in each leg.

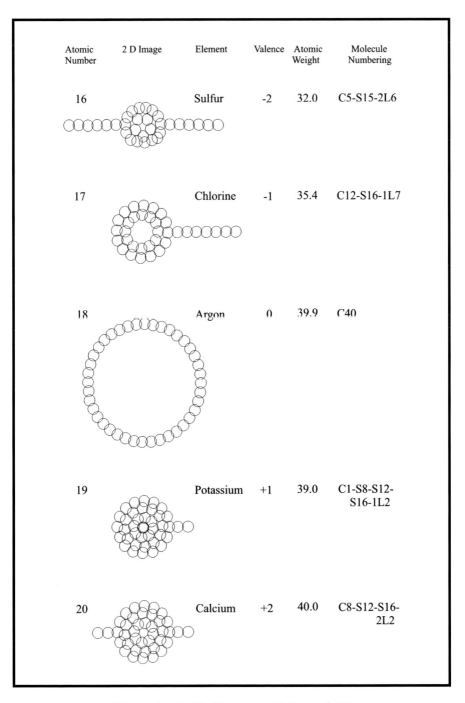

Illustration 9-20: Elements 16 through 20

SULFUR

This is a 3D image of Sulfur. Sulfur is a pale yellow in it's solid state. It melts at 119 C and boils at 444 C. Sulfur is used in making gun powder, acids, pesticides, chemical compounds and vulcanizing rubber. As you can see from the overlapping electron rings, it is a solid at room temperature.

Illustration 9-21: Sulfur

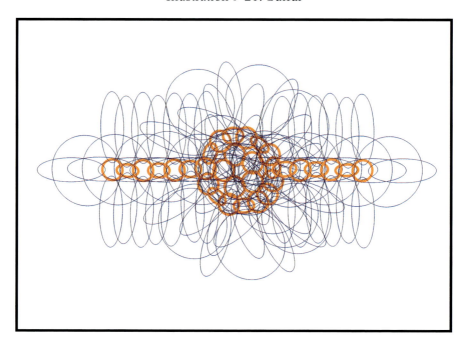

CHLORINE

This is a 3D image of Chlorine. Chlorine is a gas with a very strong smell. Chlorine is commonly found as sodium chloride or salt. Chlorine is used to make bleach and to purify water. Chlorine has a freezing point that is -101 C and a boiling point of -34 C. As you can see, Chlorine has a hollow core. It does not have many overlapping electron rings. That is why it is a gas at room temperature.

Illustration 9-22: Chlorine

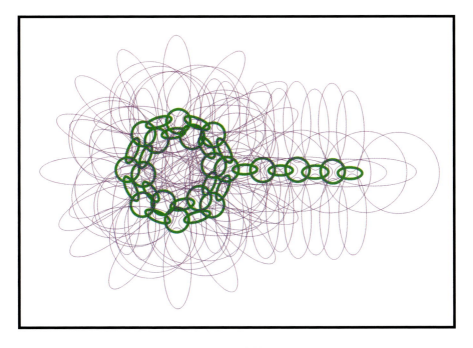

ARGON

This is a 3D image of Argon. Argon is an inert gas. It does not bond with other elements. It is colorless, odorless and tasteless. Argon makes up about 1% of the earth's atmosphere. Argon is used to fill electric lamps. Argon is also used in welding to push oxygen away from the welding arc. Argon boils at -185 C and freezes at -189 C. Argon has very few overlapping electron rings. It is a gas at room temperature. It has no legs to connect to so it is inert.

Illustration 9-23: Argon

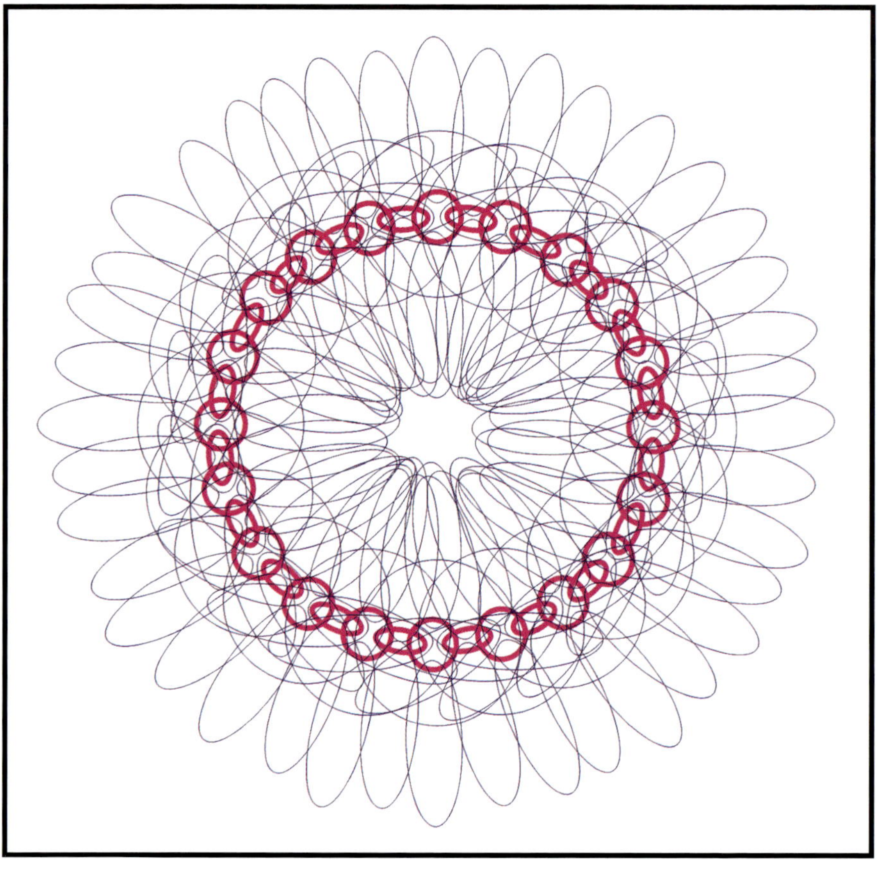

POTASSIUM

This is a 3D image of Potassium. Potassium is a silver white metal. It reacts very strongly with other elements. Potassium is all spread out throughout the earth's crust. It is the seventh most abundant element. Potassium has chemical uses and is used in fertilizer, fireworks, and gunpowder. Potassium melts at 64 C and boils at 757 C. Potassium has many overlapping electron rings. It is a solid at room temperature.

Illustration 9-24: Potassium

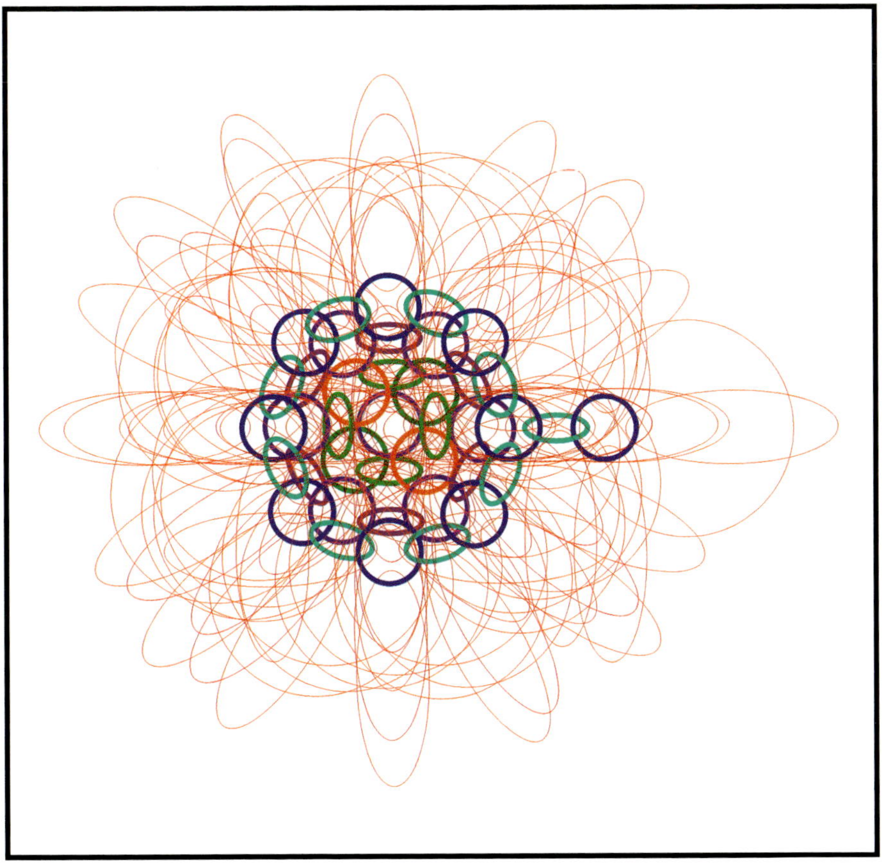

CALCIUM

This is a 3D image of Calcium. Calcium is a silver white metal. It is the fifth most abundant element. It occurs in nature as chalk, and gypsum. Calcium is the major element in bones. Calcium melts at 850 C and boils at 1440 C. Calcium easily bonds to Oxygen and Nitrogen. Calcium has many overlapping electron rings. It is a solid at room temperature.

Illustration 9-25: Calcium

Recap of molecules 16 through 20

We have moved to more complicated elements with more proton rings. The molecular chains at the center of the molecules are more complicated than the last 5 elements.

Isotopes are easy to explain. What will happen when you add a proton ring or two to the center or onto a leg? The element will act and weigh slightly different. As long as the configuration has the same number of proton rings in the center and the same number of legs, it is going to be classified as the same element.

What is different in these elements over the last 5? First, in Sulfur, the legs have 6 proton rings instead of 2 or 3 proton rings. Sulfur is very reactive and it bonds quickly with other elements. The longer legs will let it bond more easily. The second thing we do differently is we make the center of Chlorine hollow. This will give it a lower density and it will be a gas at room temperature. We gave it a longer leg because it is very reactive. The third thing is Argon, it is one big circle with no legs. It won't bond with anything. It has an atomic weight of 39.9. The next element is Potassium. It has an atomic weight of 39. Compare the two. One's a gas and the other one's a solid. The gas has a higher atomic weight than the solid. Why? With this configuration it is easy to see. With the old theory it is very difficult to explain.

Enjoy the next 5 pages, where all 20 elements are illustrated.

20 Element table

Atomic Number	Element	Valence	Atomic Weight	Molecule Numbering
1	Hydrogen	+1	1.0	C1
2	Helium	0	4.0	C4
3	Lithium	+1	6.9	C5-1L2
4	Beryllium	+2	9.0	C5-2L2
5	Boron	+3	10.8	C5-3L2
6	Carbon	+-4	12.0	C4-4L2
7	Nitrogen	-3	14.0	C5-3L3
8	Oxygen	-2	15.9	L16
9	Fluorine	-1	18.9	C12-1L7
10	Neon	0	20.0	C20
11	Sodium	+1	22.9	C1-S8-S12-1L2
12	Magnesium	+2	24.3	C5-S15-2L2
13	Aluminum	+3	26.9	C6-S15-3L2
14	Silicon	+-4	28.0	C4-S12-4L3
15	Phosphorus	-3	30.9	C1-S6-S15-3L3
16	Sulfur	-2	32.0	C5-S15-2L6
17	Chlorine	-1	35.4	C12-S16-1L7
18	Argon	0	39.9	C40
19	Potassium	+1	39.0	C1-S8-S12-S16-1L2
20	Calcium	+2	40.0	C8-S12-S16-2L2

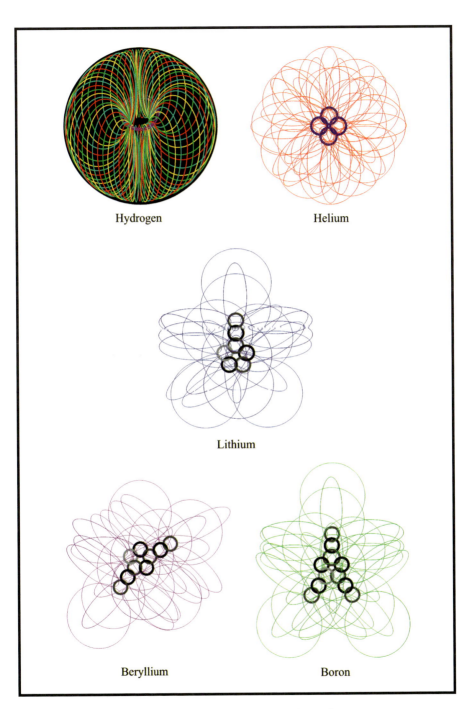

Illustration 9-26: Element 1 through 5

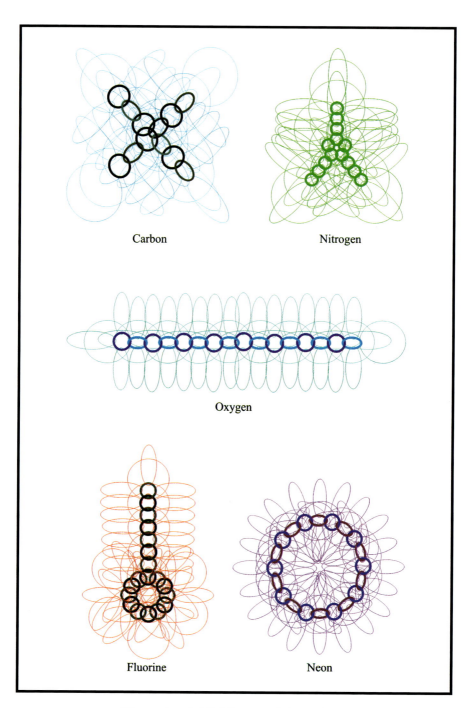

Illustration 9-27: Element 6 through 10

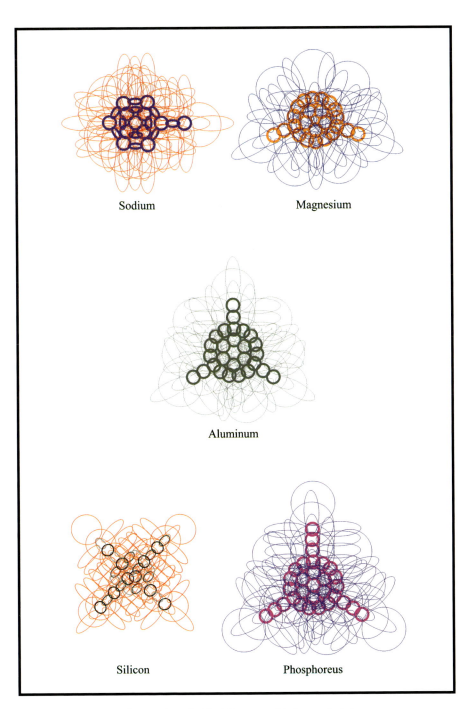

Illustration 9-28: Elements 11 through 15

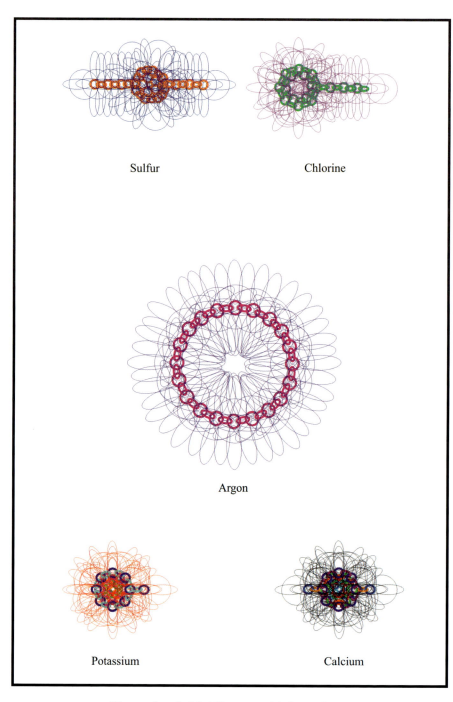

Illustration 9-29: Elements 16 through 20

COPPER

Below is a Copper molecule. Here is a departure from the first 20 elements. In this molecule I am going to show the spiral concept. This molecule is a chain of 64 proton rings. The chain spirals out from the center. This will need a new molenum. The "Sp" is the symbol for spiral. That means that Copper has a molenum of Sp64.

Illustration 9-30: Copper molecule

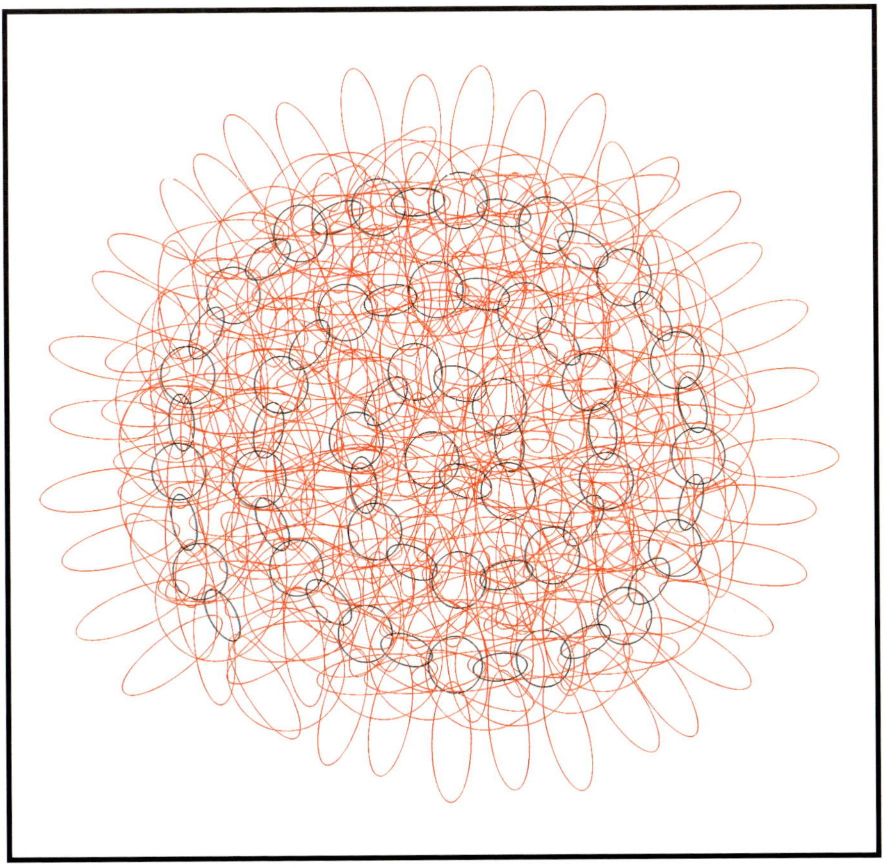

DEUTERIUM

Now we are going to move on and explain how nuclear fission works. First we are going to have to show a fissionable material. Below is Deuterium. In the past it was called an isotope of Hydrogen. The old theory model had 1 proton and 1 neutron. It was believed that when two Deuterium molecules collided it produced nuclear fusion. Fusion was a result of a proton and a neutron from 1 Deuterium atom being fused with another Deuterium molecule that had a proton and neutron. The result was a Helium molecule and a huge release of energy.

The Deuterium below is one of the isotopes that Hydrogen bombs are made of. It comes from heavy water. It is simply a chain of 2 proton rings. It has a molenum of L2.

Illustration 9-31: Deuterium

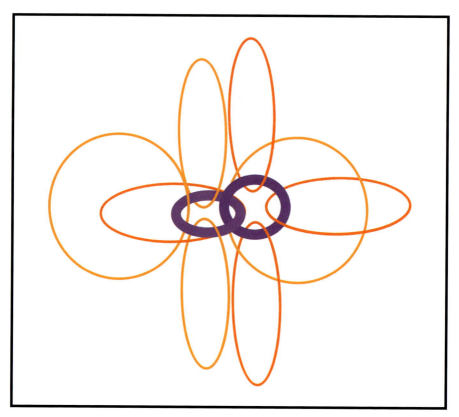

TRITIUM

This element is Tritium. It is different from Deuterium because it has 3 proton rings. However, it is similar to Deuterium as it acts like Hydrogen chemically and is used to make Hydrogen bombs. It also comes from heavy water.

Tritium is an unstable element. This means that the proton rings can be broken. Tritium has a molenum of L3.

In the following pages we will be explaining what happens when this molecule is split.

Illustration 9-32: Tritium molecule

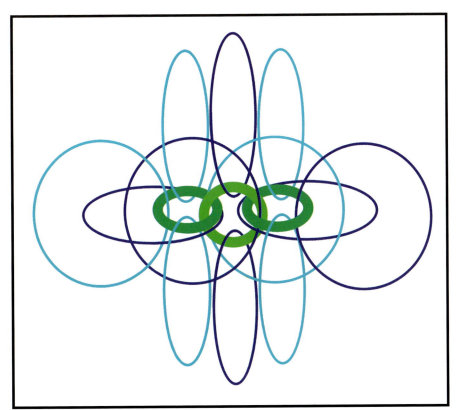

URANIUM 235

The illustration on the next page is Uranium 235. Uranium 235 is what is used to make nuclear bombs. To design this model, we used some reverse engineering. When Uranium 235 is split, it gives off energy and produces 2 elements. The 2 spiral elements are Strontium 89 and Cesium 137. Both are isotopes and radioactive. The spiral configuration is the same configuration as Copper and other metals.

Let's do some math. Strontium 89 has an atom weight of 89. Cesium 137 has an atomic weight of 137. If we add 89 and 137, the sum is 226. As we saw previously an atomic weight of 1 equals 1 proton ring. That means there are 226 total proton rings between the Cesium and the Strontium. If we do the math, there are 9 more proton rings to complete the Uranium235. The 9 missing proton rings are the energy that comes out of the bomb. We use 3 chains of 3 proton rings to connect the Strontium and Cesium together. I'm going to call the connection between spiral elements a joint. The symbol will be a J in the molenum. That means the molenum of Uranium 235 is S89-3J3-S137. By definition it's a spiral of 89 proton rings joined by 3 sets of 3 proton rings joined to a spiral of 137 proton rings. The 3 joints would be Tritium, if they were not part of this element.

Uranium has another version or isotope. It is Uranium 238 which is more stable than Uranium 235. Uranium 238 can't be used to make a bomb but is used in nuclear reactors. Uranium 238 has a molenum of S89-4J3-S137.

Uranium 235 is an extremely rare element. It is very difficult to find. It has to be mined and separated from Uranium 238. This is extremely hard to do. Uranium 235 is known as "bombs grade material". Getting the Uranium 235 is what prohibits most nations from being able to produce nuclear weapons.

This is my construction of the Uranium 235 element. It has explosives placed around it. In the following pages we are going to divide this element. Do you have any idea of where it will be split?

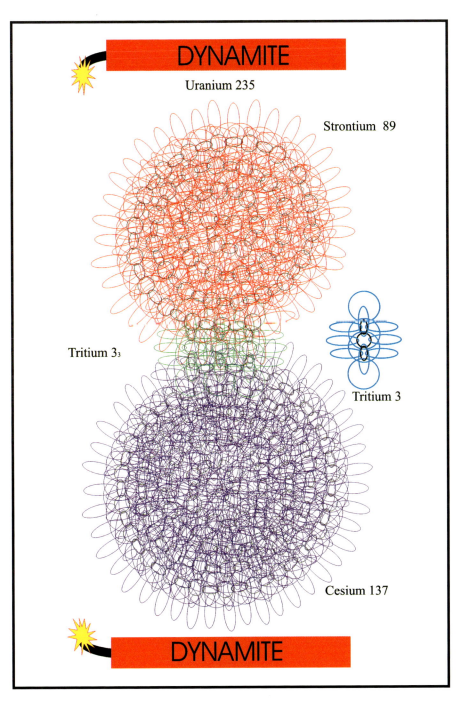

Illustration 9-33: Uranium 235

URANIUM 235 being split

The next page shows Uranium 235 being split. To make things easy to follow the Strontium 89 has red electron rings. The Cesium 137 has purple electron rings. The 3 joints with 3 proton rings have green electron rings. We will follow the colors to follow what happens. This is a nuclear explosion. These are the nuts and bolt of an atomic bomb.

To split the Uranium 235, we have to cause the explosives to detonate at exactly the same time. This will cause an implosion. The definition of an implosion is to burst inward. The implosion causes an extreme amount of pressure on the Strontium 89 and the Cesium 137. They are crushed and collide with each other. The spiral of 89 proton rings on the Strontium 89 doesn't break. The spiral on the Cesium 137 doesn't break. However, the 3 joints of 3 proton rings are crushed. This causes the proton rings of the joints to break.

To the right of the Uranium 235 is a Tritium atom. It has blue electron rings. The Tritium will soon be bombarded from particles or tadtrons coming from the atomic explosion. A nuclear explosion is the trigger for a Hydrogen explosion. The next page is the explosion.

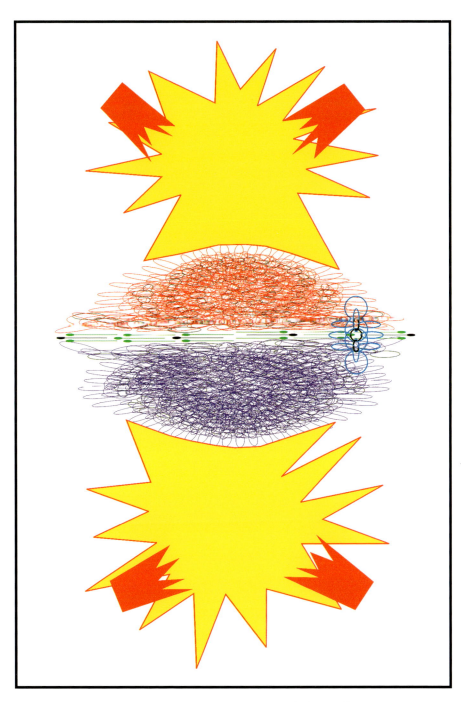

Illustration 9-34: Uranium 235 being split

A nuclear explosion
(The trigger for a Hydrogen explosion)

This is like the Hydrogen explosion in Law 8. All 9 proton rings break. All the electron rings held in place by the proton rings now separate. This produces the energy coming out of a nuclear bomb. The Cesium and the Strontium are blown in all directions. One other thing that now comes out is a neutron particle. This is an unusual particle. It has the capability of causing a proton ring to break. As the proton ring breaks, it changes its state to a neutron particle. Although I don't believe neutrons exist, I do believe neutron particles exist.

If you look at the illustration, you will notice some neutron particles have left the Uranium 235 element and have traveled right and struck the Tritium element. This causes the Tritium element to now split. It will now explode like the Hydrogen bomb in Law 8. It will release extreme amounts of energy plus some neutron particles. The neutron particles can cause other unstable proton rings to be split.

Why does the Tritium, pound for pound, release so much more energy than the Uranium? 100 % of the proton rings in the Tritium are split. Only 9 out of 235 proton rings in the Uranium are split. It's no wonder that Hydrogen bombs are more powerful than nuclear bombs.

The Hydrogen bomb explosion will now continue like it did in Law 8. We showed a Hydrogen atom being split in Law 8. This is not how it is in reality. It is Deuterium and Tritium that are actually split. They release all the energy. The only difference in Law 8 is if there were 2 or 3 proton rings instead of 1 proton ring.

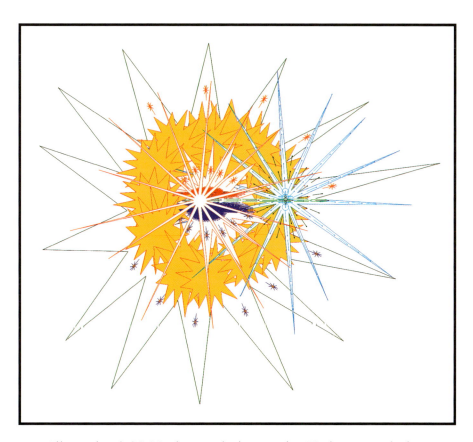

Illustration 9-35: Nuclear explosion causing Hydrogen explosion

Law 9: Closing comments

What is radioactivity? Radioactivity comes from an element that gives off radiation particles. They give off gamma rays, beta particles, alpha particles and neutron particles. All the particles are different sizes and shapes. I believe they are tadtrons in that particle state. Where does the radiation come from? I believe it comes from proton rings that are unstable. The unstable proton rings may be in a configuration that causes a bind. When the bind becomes too tight, electron rings and proton rings break and produces the radiation. The Cesium 137 and Strontium 89 have damaged proton rings. They are unstable. The damaged proton rings can come apart, this is what causes these elements to be radioactive.

Nuclear reactors use Uranium 238. It has 1 more joint than Uranium 235. It is more stable. Uranium 238 is bombarded with neutron particles. The neutron particles collide with unstable proton rings in the Uranium 238. When they collide they can cause those proton rings to break apart. In a nuclear reactor this is what happens. A Uranium atom is split. It produces some tadtrons in the neutron particle state. These neutron particles collide with another Uranium atom and cause it to split. The reaction continues on to the next atom. The reaction is controlled by substances that absorbs neutron particles. This keeps the reaction from overheating.

As you can see, I don't believe that nuclear fusion exists. I only believe that nuclear fission exists. I have opened a big can of worms. Then you are going to ask, what powers the sun? We will answer that in Law 19.

We have shown the basic ideas of the building blocks of the universe. There are many elements and more combinations of proton chains. If I were to show you more proton ring combinations, this book would get much bigger. It would detract from the main points of this theory. Therefore, I am not going to illustrate or explain any other elements in this book. The rest of the elements can be done at a later time. This is the framework of the jigsaw puzzle. I am now going to move on and start putting these building blocks or jigsaw pieces together.

Law 10

Law 10

Molecules are elements chained together by shared electron rings

Law 10: Opening comments

Once you put the framework of a jigsaw puzzle together, the rest of the pieces are easier to fit in. We now have over 20 of the elements. Next we are going to combine the elements to form several different molecules. A molecule is a combination of elements. This is going to be as easy as playing with toy building blocks.

In this chapter we are going to look at a molecule of water with the old theory. Then we are going to show a molecule of water with my new key ring elements. Let's combine more elements to make 10 more molecules. Finally we will be discussing each of the element combinations as we go along. These molecules are so facsinating that we are going to make some of them full page illustrations.

This chapter is where valence will come into play. We will be hooking things together based on their valence. We will discuss this as we show some of the molecules. After we develop several types of molecules, we will be showing some additional concepts in bonding.

Ye old water molecule

The illustration below is a molecule of water. This illustration is based on the old theory. It is based on the theory that Oxygen has 8 protons and 8 neutrons. There were 8 electrons orbiting the protons and neutrons. These 2 electrons could be shared with 2 other elements. In this example there are 2 Hydrogen atoms that share those 2 electrons. Hydrogen has a chemical symbol of H. Oxygen has a chemical symbol of 0. Since there where 2 Hydrogens and 1 Oxygen this molecule was represented as $H_2 0$.

The old belief was that the sharing of the electron held this molecule together. That means that 1 electron would figure 8 around the Oxygen and the Hydrogen. Another electron would figure 8 around the other Hydrogen atom. More or less you have 2 figure 8 electron orbits and 6 other electrons in a circular orbit. These orbits have different distances from the protons and neutrons. These orbits were called shells.

That was how it was supposed to work. I had chemistry in high school and college. I could not grasp the concept of how this produced a bond between elements. You don't see a moon doing a figure eight around two planets. Let's move onto a better molecule.

Illustration 10-1: Water the old way, H_2O

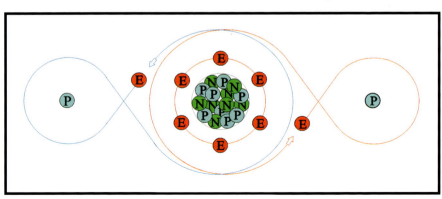

Page 178

The new water molecule

The next illustration is a water molecule with my new key ring elements that have proton chains. The Oxygen is represented with a chain of 16 dark purple proton rings. The Oxygen is bent in a semicircle. There are 2 Hydrogen elements. Each of the Hydrogen elements has a green proton ring. All of the elements have 4 red electron rings circling through each proton ring. What holds the Oxygen and Hydrogen together? A shared electron ring! The shared electron ring is represented in light blue. The shared electron ring has a larger circumference than a proton ring. The shared electron ring is not as strong as a proton ring. The shared electron ring is just like a large link in a chain. It is a weak link but it still holds.

This is where valence comes into play. Oxygen has a -2 valence. Hydrogen has a +1 valence. Oxygen can hook to 2 Hydrogens. It's that simple. Instead of electrons doing figure eights around protons and neutrons, the electrons are a weak link in a chain. The bond is easy to see.

Illustration 10-2: Water the new way, H_2O

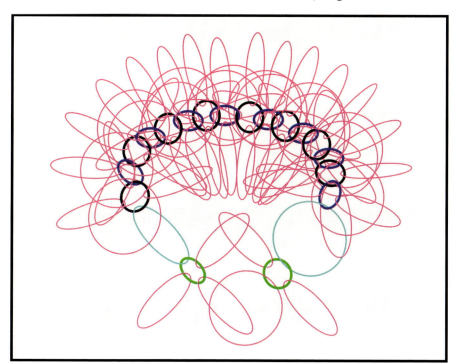

Oxygen and Ozone

This is chemistry made simple. You saw how easy it was to put Oxygen and Hydrogen together as water. The shared electron rings are known as chemical bonds. Now we are going to combine 3 Oxygen atoms to make Ozone. Then we are going to combine 2 Oxygen atoms and make O_2 or Oxygen as it occurs in our atmosphere.

On the next page are 2 molecules. The first is Ozone. The second is Oxygen. The Ozone is simply 3 Oxygen atoms chained together. Its chemical symbol is O_3. Each Oxygen atom has 16 proton rings in a chain. The chain is in a third of a circle. There are 4 red electron rings going through each proton ring. There are blue electron rings that are shared between each Oxygen atom. The shared electron rings complete the chain. They are weak links. If this molecule would break, it would break at the shared electron rings.

The second molecule is Oxygen or O_2. This is how Oxygen appears in our atmosphere. The proton rings are purple and the electron rings are red. The shared electron rings are blue. There can be 1 or more shared electron rings that hold elements together. Stronger chemical bonds may simply have more shared electron rings.

Once a molecule is formed it will have characteristics. Those characteristics will change when the temperature changes. Molecules will be either a solid, a liquid or a gas, based on its temperature. We will cover this in Law 11. Both Ozone and the O_2 are a gas at room temperature. Density also plays a role in molecules being a solid, a liquid, or a gas. There are very few overlapping electron rings so the density of both molecules will be low. That's why they are a gas. Which one looks lighter? Which one will go higher in the atmosphere? Ozone does! Notice the big space in the center of the Ozone. Ozone is less dense than O_2 so it's lighter and it will go higher in the atmosphere.

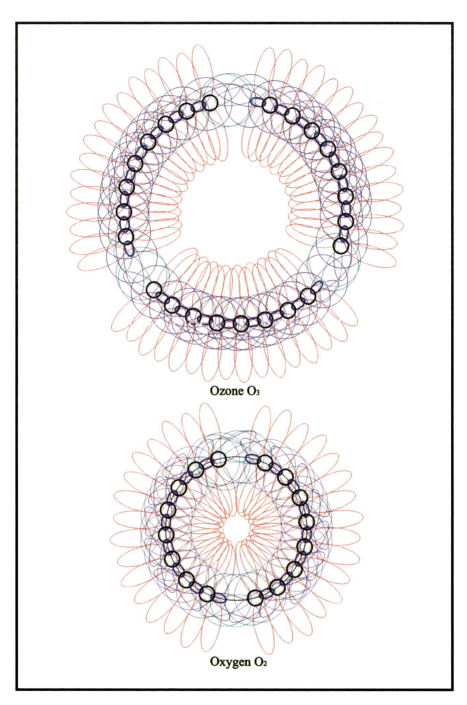

Illustration 10-3: Ozone and Oxygen molecules

Carbon Dioxide, CO_2

On the next page is a molecule of Carbon Dioxide. It's chemically known as CO_2. The molecule consists of 1 Carbon atom and 2 Oxygen atoms. The Oxygen atoms are above and below the Carbon atom. Each Oxygen atom has 16 dark purple proton rings. The Carbon atom has black and gray proton rings. All of the electron rings are red. The shared electron rings are blue.

What did we do? The Carbon has a +-4 valence. It has 4 connection points. Oxygen has a -2 valence. It has 2 connection points. We hooked 1 Oxygen to 2 connection points on the Carbon. Then we hooked the other Oxygen to the remaining connection points on the Carbon. We used 4 shared electron rings. This is a Carbon Dioxide molecule.

Carbon Dioxide is a gas at room temperature. Look at the molecule and take a guess at its density. Is it more dense than O_2? Yes, it is. Notice the space or gap between the Carbon and the Oxygen. This space makes it less dense so it can be a gas at room temperature. If we changed the size of the electron rings, would this molecule have different characteristics? Yes, it would. When we deal with temperature in Law 11, we will discuss the size of the electron rings.

How is Carbon Dioxide formed? When a substance made of Carbon is burned, it will produce Carbon Dioxide. What happens when something is burned? The 0_2 breaks and links up with Carbon. Shared electron rings are broken and reformed. It's similar to playing with building blocks. You can take them apart and put them back together.

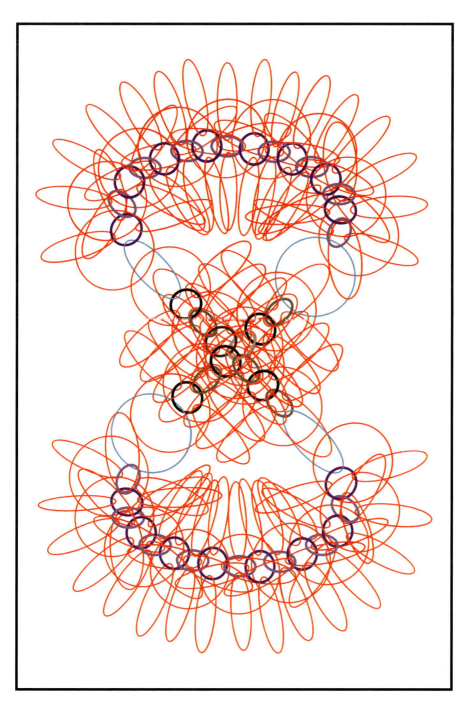

Illustration 10-4: Carbon Dioxide, CO_2

Diamond, C_4

The illustration below is a Diamond. There are 4 Carbons that are hooked together with 8 shared electron rings. This is the hardest natural substance known. A real Diamond will continue chaining Carbons until there are no more. A Diamond is really just 1 big molecule. There is a square hole in the middle. It would appear to have a low density. But remember the molecule is one big chain of many Carbon atoms. When they are hooked in and overlap in 3 dimensions this space would be less. A Diamond is a solid at room temperature.

A Diamond is transparent. Light will pass through it. Light seems to like to go through square holes. We will discuss this in Law 16.

Illustration 10-5: Diamond, C_4

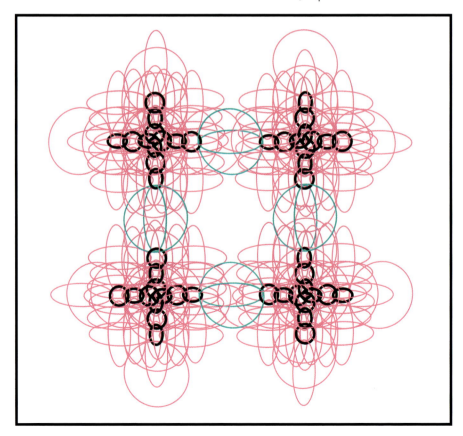

Ammonia, NH$_3$

The illustration below is Ammonia. Its chemical formula is NH$_3$. The Nitrogen has green proton rings. Everything else is the same as before. This is simply a Nitrogen atom hooking to 3 Hydrogen atoms. Shared electron rings hold them in place. The density is lighter than air. There are very few overlapping rings. Ammonia is a colorless gas that has a bad smell and will burn the inside of your nose.

Illustration 10-6: Ammonia, NH$_3$

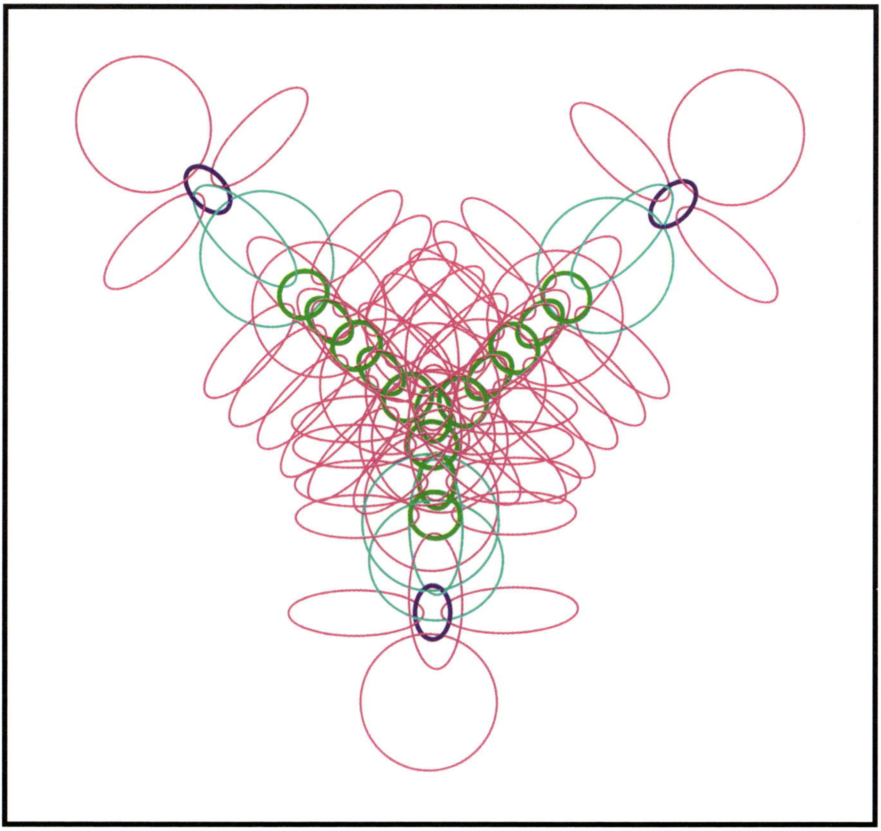

Nitrous Oxide, N$_2$O

The next page contains an illustration of Nitrous Oxide. It has a chemical formula of N$_2$O. In this drawing, we have 1 Oxygen atom between 2 Nitrogen atoms. This molecule is not complete. Each of the Nitrogen atoms have 2 more connectors. Would they be available to hook to something else? If this is the way the molecule is, then yes they could hook to another atom. But in reality these 2 ends probably hook to each other. This is almost impossible to illustrate. On this illustration you will have to finish it with your imagination. Bend the Oxygen and hook the ends together. It is now a complete molecule.

Nitrous Oxide would be round in shape. It is heavier than air. Nitrous Oxide is also known as laughing gas. It has been used as an anesthetic. Nitrous Oxide also works better in combustion than air. Better combustion means it burns better.

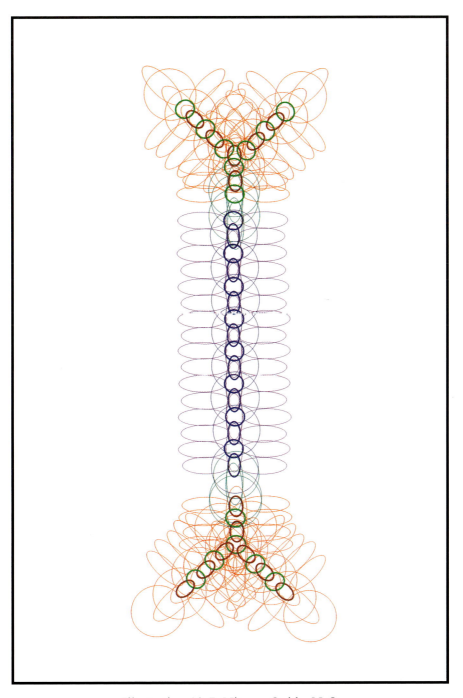

Illustration 10-7: Nitrous Oxide, N₂O

Pure Nitrogen, N_2

The next page contains an illustration of Pure Nitrogen. This is how it exists in our atmosphere. I have shown only 1 set of shared electron rings. I think that there should actually be 3 sets of shared electron rings. Each of the ends of the proton rings should be connected. It would be very difficult to show this in an illustration. You would have to fill in the gaps with your imagination. Just imagine the longer ends of the Nitrogen bending and hooking together. It would almost be like folding a piece of paper. Image the next page curled and then touching the ends. What would you have? It would be a sphere with a hollow center. This is going to give it a density about the same as Oxygen. The N_2 is then a gas.

N_2 is incombustible. It won't burn. About the only thing that reacts with N_2 are plants. Some plants can take the Nitrogen, out of the air and put it in the soil. Nitrogen in this molecule makes up about 78 % of our atmosphere.

There are a lot of molecules that use Nitrogen. Nitrogen will always hook in combinations of 3. That's how it works. It's just like building blocks that have 3 connection points.

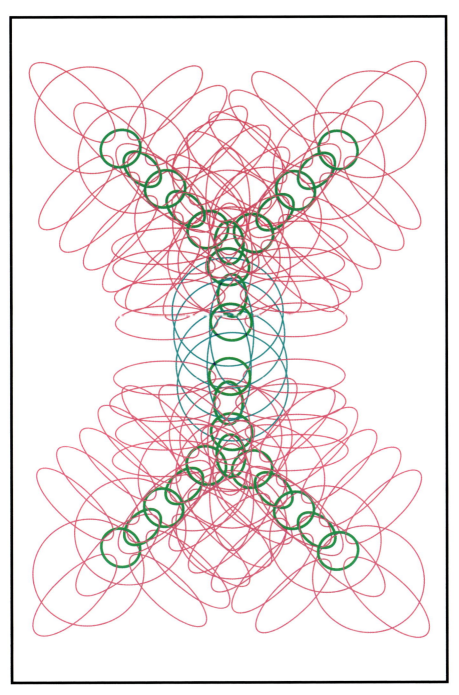

Illustration 10-8: Nitrogen, N_2

Silica or Silicon Dioxide, SiO_2

We are now going to move onto a molecule with Silicon. The Silicon atom has black proton rings. We still have red electron rings. We have hooked an Oxygen atom above and below the Silicon atom. We have used bluish-green electron rings to cause the chemical bond.

This molecule is a lot like Carbon Dioxide. There is one big difference between Carbon Dioxide and Silicon Dioxide. Carbon Dioxide is a gas at room temperature. Silicon Dioxide is a solid at room temperature. What causes this difference? One thing is the Silicon is more dense at the core. The second thing is the Oxygen is closer to the Silicon. This causes overlapping electron rings. That causes the density to be higher.

Silicon Dioxide is also known as Silica or sand. This molecule is found in 95 % of the rocks in the earth's crust. Silica is very resistance to change. It is a very hard substance as well. Carbon Dioxide floats off as a gas. Silica stays together unless it is heated to a high temperature. What's the difference? Silica may form yet another chain. So far, we have had a chain between proton rings and electron rings. The next chain is just like handcuffs. Look at the Silica. It looks like handcuffs. How would you hook handcuffs together? Unhook the end of one handcuff. Then slide the end through the hole in the other handcuff. Close the open handcuff. You now have a chain of handcuffs. You could hook multiple handcuffs to make several combinations of chains.

Now apply the handcuff chain to the Silica. This will account for how Silica acts. Silica forms quartz and several other crystalline structures. The handcuff chain can form all these structures. These handcuff chains will also cause the density of the Silica to be much higher. This will also account for why it is a solid. Silica has lots of properties. I don't have room in this book to cover them. You will have to do your own research to see what they are.

Silicon can also combine into a regular chain of proton rings and shared electron rings. I am going to show that when we show Glass.

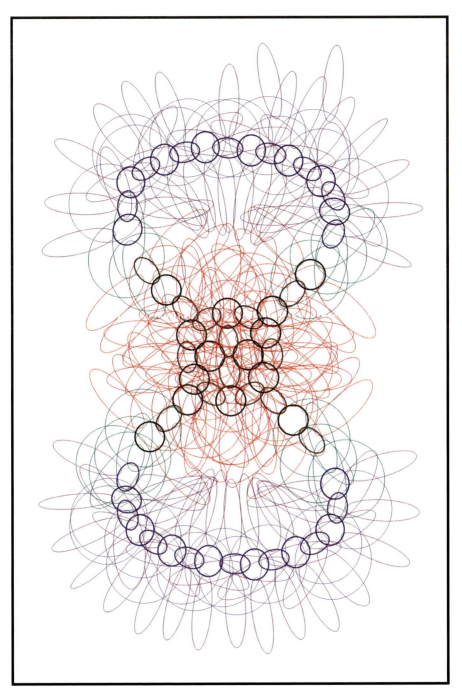

Illustration 10-9: Silicon Dioxide, SiO$_2$

Glass, Si_4O_4

This molecule is the inner working of Glass. You can tell which is Silicon and which is Oxygen. The shared electron rings are bluish-green. These can make one big molecule like a Diamond. There is a square hole. Glass is transparent. Light likes to go through square holes.

Other atoms are also in Glass. Aluminum is one of them. Aluminum could be at the end of a Silicon atom. Oxygen could round off the ends of the Glass. With this big square hole, you can still do the handcuff chains. Glass has many variations. There are an almost infinite number of ways to put molecules together.

Illustration 10-10: Silicon Oxide, Si_4O_4

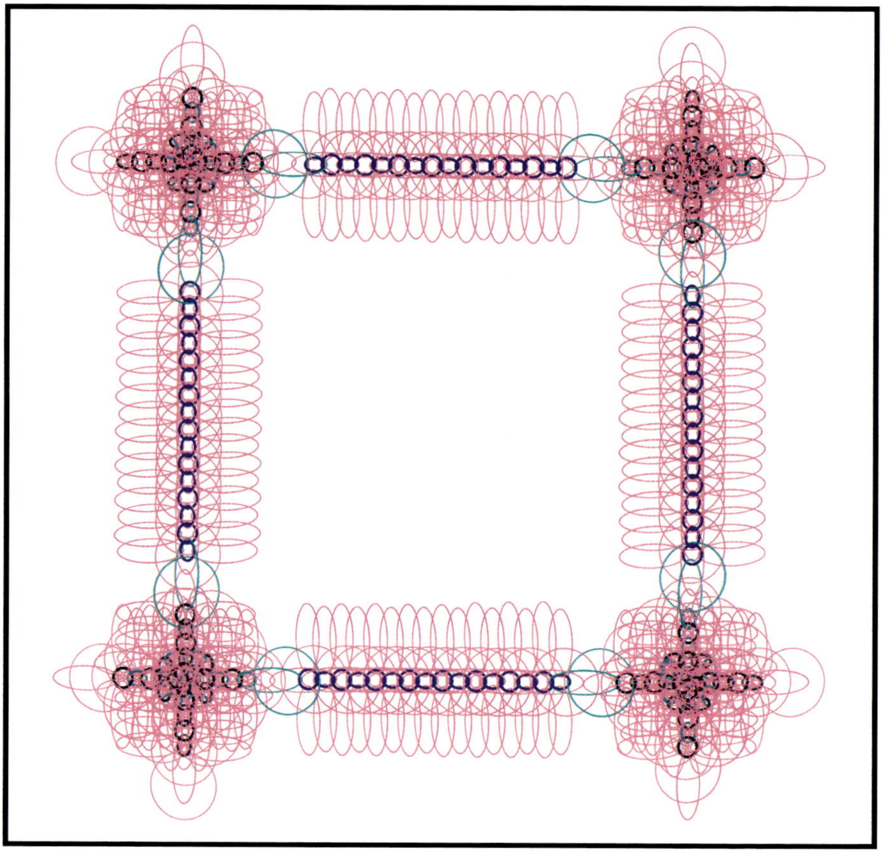

Combined Copper

There is one more bonding that I would like to discuss. It's metal bonding. I think most of the metals are a spiral chain of proton rings. How are they held together? Inner twining of electron rings. Have you ever had 2 coil springs stuck together? They are very hard to get apart. This is how the metal molecules bond.

I have 4 Copper molecules. Each has a spiral of black proton rings. The red rings are the electron rings. The electron rings of each copper ring are meshing into the next copper electron rings. This meshing allows for bending in the metals. They are not hard bonds like shared electron rings. There is a round hole. Electricity likes to go through round holes.

Illustration 10-11: Copper molecules

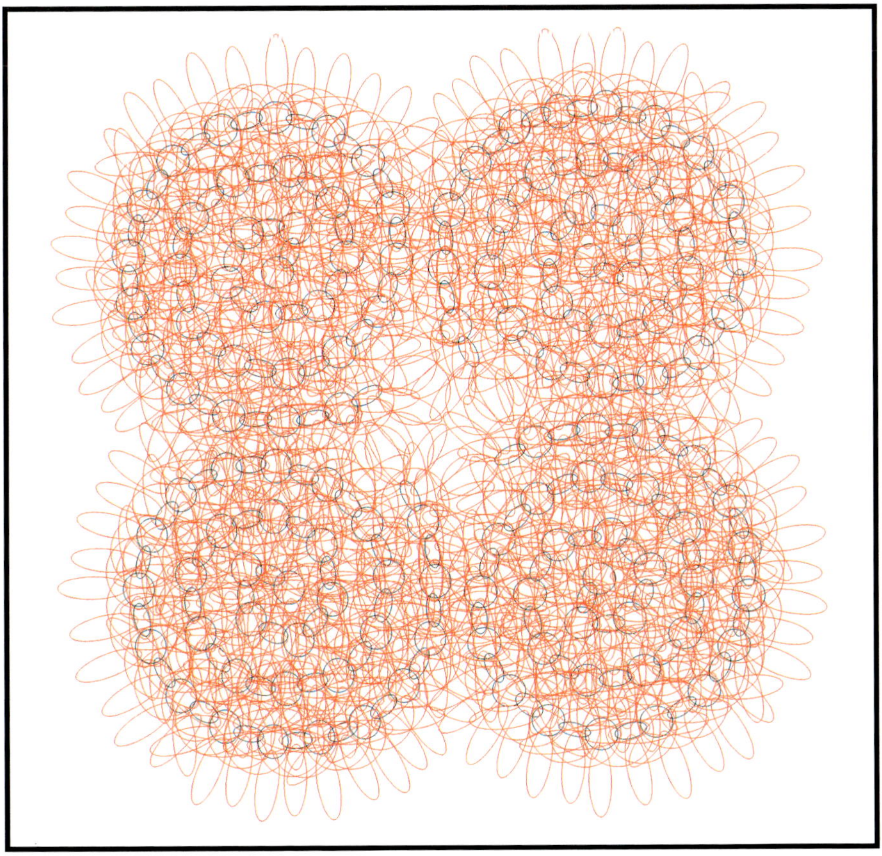

Law 10: Closing comments

I hope you liked this chapter. It ties up a lot of loose ends. Most molecules are bonded together by shared electron rings. Some molecules are bonded together with a handcuff chain. Some molecules are meshed together. This is a big section of the jigsaw puzzle.

When you combine the different types of bonding with the known elements, you arrive at an almost infinite number of combinations. There are believed to be over a million different combinations of carbon molecules! There are only 10 molecules shown in this law. It doesn't even scratch the surface of what is out there.

What have I changed over what was believed in the past? Actually very little. I have changed the part of the atom that no one could see. I took the proton and neutrons out and replaced them with a key ring chain of proton rings. These chains give each element its characteristics. The characteristics are the same. I have not changed them. All the work in chemical bonding is the same as before. The next chapter will talk about how those characteristics are influenced by temperature.

Law 11

Law 11

Heat and cold is caused by the size and speed of an atom's electron rings

Law 11: Opening comments

The jigsaw puzzle is getting easier. The framework of the puzzle has been built. Heat and cold is another section. It's time to put these pieces together.

In the past heat and cold were believed to be caused by the protons, neutrons and electrons all vibrating. As an element vibrated faster it was considered to be hotter. When the vibrations slowed down it was colder. Absolute zero was considered to be when a molecule didn't vibrate. This old theory is not going to fly anymore.

In the last three laws we defined atoms, elements and molecules. We could visually see the characteristics of elements and molecules. We could see how elements could connect to other elements. We could see how they act after they were connected. We could see this based on their shape. Heat and cold do change the characteristics of elements and molecules. Could temperature change the shape of an element or a molecule? Is that an answer? We will soon see.

The size of the electron ring changes the shape of an atom

The Hydrogen atom has the shape of a donut. It is also shaped like an inner tube. I am talking about an inner tube that fits inside of a tire on a car or a truck. Let's take an inner tube out of a tire. First, we will make sure the tube has no air in it. What does the tube do? It lays flat. Next put a little air in it. What does the tube do? It takes the shape of a donut. It still lays flat. It has a little bounce to it and it could be rolled. Next add some more air to it. What happens? It doesn't sit as flat. It's easy to roll. It's easier to bounce. Now add more air. What happens? The tube doesn't easily sit flat. It will bounce, if it is rolled. If you bounce the tube it will bounce and bounce and bounce. Inflating an inner tube is very similar to changing the temperature of an atom. That is where we are going.

On the next page is an illustration of 2 Hydrogen atoms. The top one is a hot Hydrogen atom. The bottom one is a cold Hydrogen atom. Both atoms have a single purple proton ring. Both atoms have 4 red electron rings. The proton ring is the same size on both atoms. The proton ring overlaps about 3 times.

The hot Hydrogen has large electron rings. The electron rings are 3 times the circumference of the proton ring. The electron rings are spread out and there is a small gap between the head and the tail. The cold Hydrogen has electron rings in a circumference that is the same size as the proton ring. The size and the shape of the 2 atoms are different.

Heat and cold is that simple. When an atom is heated up, the circumference and speed of its electron rings increase. The more it is heated, the larger the atom gets. Its characteristics will change. It will go from laying flat, to rolling and then bouncing. It's just like putting more air in an inner tube. When an atom is cooled, the circumference and speed of its electron rings decrease. The more it is cooled, the smaller the atom gets. Its characteristics will change. It will go from bouncing, to rolling, and then laying flat.

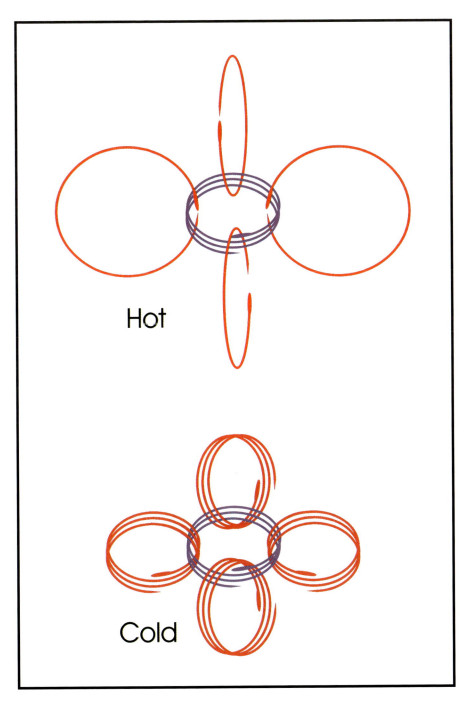

Illustration 11-1: Hot and cold Hydrogen

Steam, water and ice

On the next page are 3 molecules of water or H_2O. The top molecule is water when it's a gas. The second illustration is water when it's a liquid. The third illustration is when water is a solid. The common names for H_2O in these states are steam, water, and ice. What's the difference between them? Just the circumference of the electron rings.

All the proton rings in all atoms are the same size. Steam has electron rings with a circumference about 6 times the size of its proton rings. Water has electron rings about 4 times the size of its proton rings. Ice has electron rings about twice as big as its proton rings. The size of the electron rings will account for how water acts when it is heated and cooled.

When water is heated, the electron rings get bigger. The density of the molecule goes down, it will float or bounce around. When that water is cooled it will become a liquid. The water will curl in a ball. It will be round. It will roll. When the water is cooled more, it becomes a solid. It will straighten out, become rigid and lay flat. When water freezes or turns to ice, it expands. The expansion is due to the Oxygen uncurling from a ball.

This is how all elements and molecules act with temperature change. As the temperature rises, outside of the elements or molecules change. When they are cold, they lay flat and they are solid. When they get round, they roll like a ball and become a liquid. When they get bouncy, they become a gas. It's not too complicated.

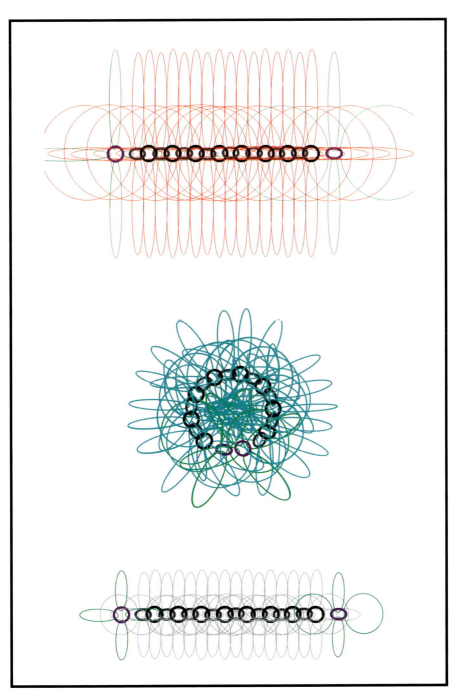

Illustration 11-2: Steam, water and ice

Heat transfer

Heat transfer is what happens when 2 substances that have a different temperature come in contact with each other. They exchange temperature. The hot substance will cool and the cold substance will warm up. An example would be mixing a gallon of water at 100 degrees with a gallon of water at 50 degrees. What would be the result? You would get 2 gallons of water at 75 degrees. What are the mechanical workings behind heat transfer?

On the next page I have 6 Hydrogen atoms. The proton rings are purple. There are only 4 electron rings. The Hydrogen atoms are paired off. The first pair at the top has a hot Hydrogen on the left and a cold Hydrogen on the right. The second pair of Hydrogen atoms is in the middle. We have moved them closer together. The atom on the left has an electron ring touching the proton ring on right. This starts a spin down on the left atom. It will cause a spin up on the right. The third pair of atoms are at the bottom. All electron rings are now the same size. Both atoms are now warm. They are not hot or cold.

What happened? The hot atom had an electron ring that was spinning faster with a larger circumference. When the hot atom came in contact with the cold atom, the speed and size of the electron rings were transferred to the cold atom. The electron rings on the cold atom spun faster and increased in circumference. As the cold atom warmed, the hot atom cooled. The hot atom's electron rings slowed and decreased in circumference. The heat transfer happens until both atoms are the same temperature. This is the mechanism behind heat transfer.

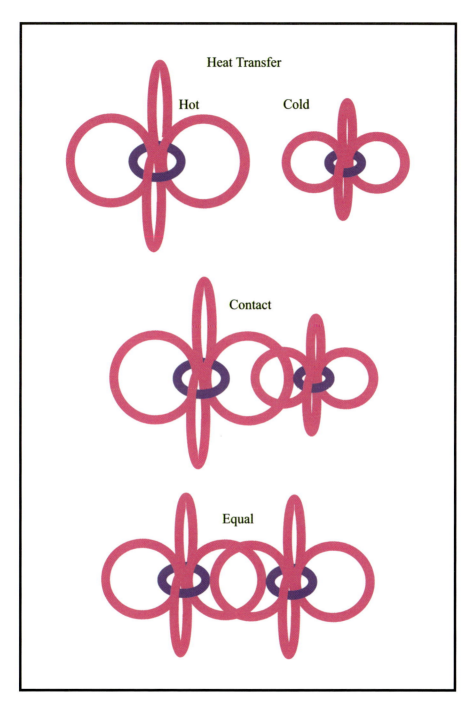

Illustration 11-3: Heat transfer

Law 11: Closing comments

This was an easy section of the puzzle. Temperature plays a huge roll in the characteristic of elements and molecules. This is where you get to do a little back tracking. I have tried to put an illustration on the same page, or on the opposite page, of my writing. This gives you a visualization of what I am trying to explain. To give an example of temperature changes in all the elements and molecules I have in this book would result in thousands of pages.

Now go back and look at each element and each molecule. Increase the size of the electron rings with your mind. What will you notice? They all get rounder. What happens when they get rounder? They roll. What happens when they roll? They become a liquid. Increase the size of the electron rings even more. What happens? They get bouncy. What happens when they get bouncy? They become a gas.

When molecules get cold, the electron rings get smaller. Have you ever had 2 coil springs tangled up together? It's a mess. They are very hard to get apart. When items become solids, there will be a lot of electron rings that will be tangled and meshed together. This is what will allow them to stick to each other. If elements are bonded, they will be solids. If elements are tangled, they may be just as dense. The tangled electron rings can also account for why some solids have the ability to bend. Look at the Copper atom. It will be a tangled mesh of electrons. If we had iron or steel it would be similar. To make iron and steel harder it is heated and then pounded with a hammer. What does the heating and pounding do? It causes the electron rings to expand and then the pounding causes more of them to be tangled. This causes the iron or steel to become much harder.

Liquids also have an adhesion characteristic called viscocity, which means they like to stick together. Look at maple syrup, it is very sticky, it is slow when it is poured. What accounts for this? Tangling and meshing of electron rings. Even though liquids are in a ball, the electron rings will overlap and cause some of the balls to stick to each other. This will explain surface tension. The surface tension characteristic is the same as adhesion. The difference is how it is viewed. An example of surface tension would be a water drop on top of a waxed car. What happens? The water curls up in a ball rather than laying out flat. It beads. This is caused by the overlapped and tangled electron rings.

I would like to discuss a characteristic of gases. When you put a gas in a closed container, you have some pressure. Pressure is defined as a unit of force exerted on a surface. What causes this exerted force? It is caused by a couple of things. One is the speed and direction of the electron rings. When molecules collide, some of the colliding electron rings will be going in opposite directions. This will cause the molecules to bounce apart. The second thing is an open edge on some electron rings. The open edge comes from a gap between the electron ring and the proton ring. When this open end comes in contact with another molecule, it will cause a bouncy repelling effect. The movement and collision of electron rings will cause pressure in a gas.

Another thing that will happen with changes in temperature is molecular bonding. Molecules that make bonds at a low temperature, may break at a high temperature. Some bonds may not occur unless the temperature is high. Why would molecular bonding change? The shape of the elements will change with temperature. The size of the shared electron ring will change with temperature. Temperature will have a big effect on how molecules bond. Go back and look at all the molecules that we bonded in Law 11. Increase the size of the electron rings and see how different pressures will be placed on the bond.

Heat transfer among elements will be different. Some elements may easily transfer heat. Some elements may have a hard time transferring heat. Heat transfer will depend on how the electron rings align and interact with one another.

The speed of electron rings are unknown. The only speed that is relative is that as temperature increases so does the speed of electron rings.

Law 12

Law 12

All molecules have energy streams that travel through their centers

Law 12: Opening comments

We have tried to piece things together like a jigsaw puzzle. Gravity is a huge section of that puzzle. There are 4 parts to gravity. This is part 1 of the gravity section.

The goal here is to define gravity. Laws 12, 13, 14, and 19 will provide a new way of looking at how gravity works. We have been told that gravity is a force of nature that pulls 2 masses together. It is also defined as the attraction of masses for one another. What proof do we have that bodies attract each other? What if it is something else? What if it can be explained in a better way?

Let's look at the old way of explaining gravity. For gravity to be able to pull or attract you must have 2 things. First, each of the 2 bodies or masses must know the other exists. They must be able to see one other. Second, they must some how attract or pull each other. Roping a steer would be an example of seeing and pulling. First you would have to see the steer. Then you would have to throw the rope around the steer's neck. Then you would have to pull on the rope to move the steer. Is this how gravity works? No, it isn't. Prior to roping the steer something is coming from the steer. What? Reflected light. That's how we see him. This is the only thing we know that is coming from the steer. Remember this.

Another example of seeing and pulling is a frog catching a fly. For a frog to be able to catch a fly he must first see the fly. We saw earlier that light can be reflected from matter. So what a frog really sees is light being reflected from the fly. The frog's eyes detect the light coming from the fly. Then the frog flicks out his tongue and catches the fly. We can see again, that a stream of energy (light) is already coming from the fly to the frog's eyes. Remember this stream of energy.

Let's use another example that doesn't see. If a bat is blind, then how does a bat catch a fly? I think most people know that bats send out sound waves. These sound waves bounce off the fly. The bats have very sensitive ears that pick up these sound waves coming from the fly. The bat then catches the fly. Once again we can see a stream of energy (sound) coming back from the fly going to the bat's ears.

Another example is radar. How does radar work? It's the same principle as the bat. Radar waves are sent out. The radar waves are deflected from an airplane. The radar detector picks up the deflected radar. Once again we see a stream of energy (radar) coming back from an airplane to the radar detector.

How is all this going to apply to 2 molecules floating in space? How are they going to see each other? Is each molecule going to have an eye that can see light reflected? No, or else gravity would not work in the dark. Does each molecule send out sound waves? No, that wouldn't be practical, they would have to have ears to listen. Does each molecule send out radar waves? No, that wouldn't be practical either, they would have to have a radar detector to pick up the radar waves being reflected. So how do 2 molecules in space see each other? They don't. If they don't see each other, then how do they pull or attract each other together. They don't pull or attract, they do something else. There is another answer. Let's look at it.

Vacuum cleaner effect

The illustration below is a vacuum cleaner. How does it work? A vacuum cleaner has a motor inside it that turns a fan. When the fan turns, air goes through the fan. A lower pressure is produced. The air, around the end of the vacuum cleaner hose, follows the air ahead of it into the vacuum cleaner hose. The higher pressure follows the lower pressure. All the air around the end of the hose is affected.

Streams of air go into the hose. That's shown by the arrows. There are feathers floating in the air. The streams of air go around and through the feathers. When the air goes around and through the feathers, they are pushed towards the end of the vacuum cleaner hose. From a distance it would look like the feathers are being attracted or pulled into the hose.

Is this how gravity works? Is a stream of energy already going through the molecule? Is the molecule like the end of a vacuum hose? Let's turn the page and see.

Illustration 12-1: Vacuum cleaner

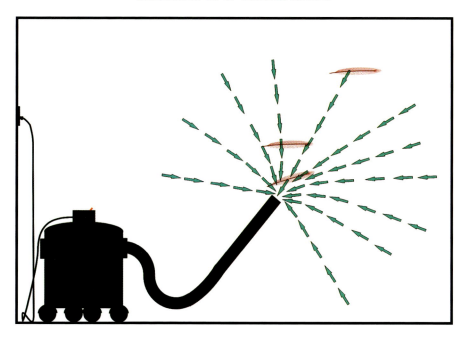

A stream of energy goes through the center of all atoms

This is the answer to gravity. Each key ring atom has a hole in the center. The hole is in the proton ring. All the electron rings have a circular orbit. The circular orbit has a rotation that goes in on one side and circles out on the other. This Hydrogen atom has the electron rings rotating to the inside. This inside rotation is like the fan inside the vacuum cleaner. The rotation is what draws or attracts tadtrons that will become gravity and then produce gravity streams.

This is the mechanical working of gravity. A Hydrogen molecule sits there with rotating electron rings. A tadtron is flying around in space. It has an energy state that has the following job, find a molecule and travel through its center. When this tadtron finds a molecule, it straightens its tail to point to the molecule it is going to. This tadtron will now slow down in speed. The straight tail is a directional pointer. This tadtron knows where it is going. Its job is to go through the center of the molecule. Another tadtron in an energy state has the job of finding a tadtron that knows where its going. The straight tail is the key. This tadtron slows down and gets in a line behind the tadtron that has a straight tail. It straightens its tail in the direction of the tadtron in front of it. It then continues in that direction.
A long line of tadtrons form. They are now a gravity stream.

In the illustration on the next page, tadtron 1 is going through the center of a hydrogen molecule. Tadtron 2 is following tadtron 1. Tadtron 3 has found tadtron 2 and is slowing down and straightening out its tail. Tadtron 4 is looking for a tadtron that, "Knows where it is going". When it meets tadtron 3 it will straighten its tail and follow the gravity stream. This is very similar to the air going into a vacuum cleaner hose. The air follows air. In this example the tadtrons follow tadtrons.

I am going to call the tadtrons in the gravity streams gravitons. There may be millions of gravitons going through each atom per second. It could be more gravitons. There could be less.

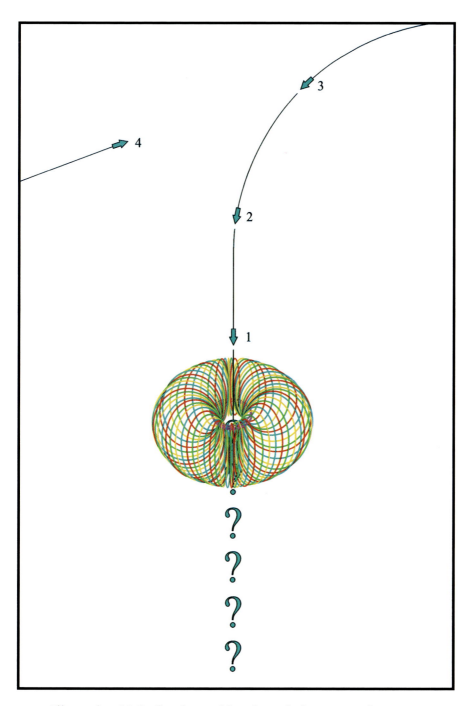

Illustration 12-2: Gravity pushing through the center of an atom

Law 12: Closing comments

This is part 1 of the gravity section of the jigsaw puzzle. The idea here was to show that all molecules have streams of energy that travel through their center. This is as easy to understand as air going into a vacuum cleaner hose. One thing to note is that the gravity only goes through 1 side.

What is the configuration of a tadtron that is in the graviton state? The head will be very small. The tail will be very long and straight. The graviton will be very thin and very small. We saw earlier that gravity travels through just about everything. It seems to be unaffected by matter. We will show gravitons as arrows in all the rest of the gravity illustrations.

This law covers energy going through the center of all atoms. There are a couple of things we haven't covered. What is the effect of these gravity streams on other atoms? How long are the streams? What happens to the gravity after it exits a molecule? We will answer these questions in the next several laws.

Law 13

Law 13

**Gravity is an energy stream that pushes
all molecules in its path**

Law 13: Opening comments

This is part 2 of the gravity section of the jigsaw puzzle. Gravity has always been the odd man out. No one has ever been able to give an easy to understand definition of gravity. At the end of this law, you will understand gravity like you have never ever understood it before.

Gravity was defined as the attraction bodies have for one another. The goal of this law is to explain this alleged attraction. What we will be showing you is not an attraction. It is something entirely different. In Law 12 we saw that all atoms have a stream of energy that go through their centers. In this law we will be showing the effect of this stream of energy on all the surrounding matter.

Here is an easy example to follow. We will use the wind as an example of energy and we will use my body as the mass. First, let's say the wind is blowing 50 mph. Second, let's say there is a tree down wind. Third, let's say I am on a skateboard and there is nothing but flat smooth concrete between the tree and my body. The wind is directly behind me. As the wind blows, I will start to move towards the tree. I will start slowly but I will accelerate up to 50 mph, the same speed as the wind. Then I will slam into the tree. To someone who was watching, it might appear that the tree attracted me or pulled me right into it. Is this what happened? No, I think we all know that the wind pushed my body into the tree. In this example of nature, two bodies didn't attract each other. I was pushed by the wind. I was also pushed in the direction that the wind was blowing.

A drain effects fish in a pool

On the next page is an illustration of a swimming pool. There are 3 fish in the pool. There is a drain at the bottom of the pool. The drain has been opened. Water is traveling out of the drain. What is the affect on the fish? The fish start to move towards the drain. Which fish will be the most affected? The fish that is closest to the drain will feel the most pull.

If a person watched this from a distance, it would look like the fish are being pulled into the drain. The fish would even feel a pull. Are the fish pulled into the drain? No, they aren't. They are caught in the stream of water that is moving towards the drain. The streams of water push the fish towards the drain. I have used arrows to show the streams of water. The fish closest to the drain has the most arrows going through him. That fish will be pushed the most. The fish in the middle will have 3 arrows going through him. He will feel a much smaller push than the first fish. The fish at the top only has 1 arrow going through him. He will only feel a slight push.

Another example of being pushed by water would be walking into a river. If you walk into the river, where the water is 1 inch deep, you will feel a very slight push by the current. Next, take another step and you will be 1 foot deep in the river. You will feel a much stronger push, but you will not go down the river. Take one more step and you will be 2 feet deep in the river. You will feel a much stronger push. Now take another step and you will be 3 feet deep in the water. The water will be pushing you very hard at this point. Take another step. At this point the push of the current is too strong. You will now be pushed downstream. You will accelerate to the speed of the river. In this example you can see as more water from the river hits your body, the more you will be pushed.

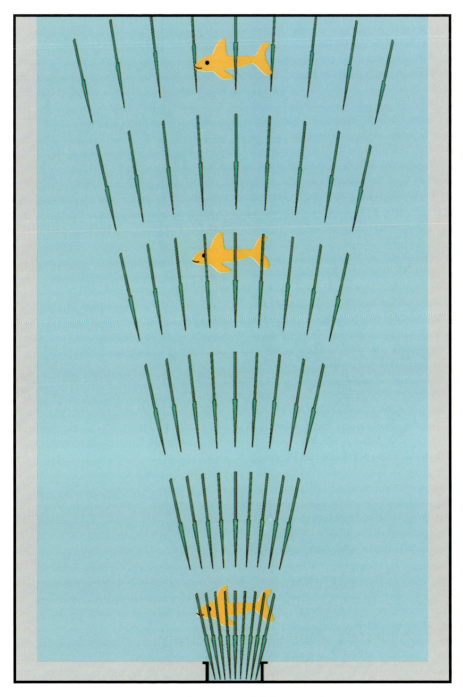

Illustration 13-1: Fish in a pool

The air going to a vacuum cleaner hose pushes feathers in that direction

Before we look at the illustration, get a feather and blow on it. It is pushed by the air from your mouth. The closer the feather is to your mouth the more effect you have on it. More air equals more push.

In the next illustration we have the vacuum cleaner. Inside the vacuum cleaner is a fan that turns and causes air to move into the hose. The air moves into the hose. It will come out of an opening in the vacuum cleaner. All the air around the end of the hose will be effected. Air will follow air. There will be streams of air going in the hose. There are some blue arrows that show the streams of air.

There are 3 feathers above the end of the hose. The feather closest to the end of the hose has 5 blue arrows going through it. The 2nd feather is farther away from the end of the hose. It has 3 blue arrows going through it. The 3rd feather is farther away from the end of the hose. It has only 1 blue arrow going through it. The blue arrows are a simulation of air streams. What will be the result of these air streams flowing by the feathers? The feathers will be pushed. If you have more streams, you will have more push. The feather closest to the end of the hose will get pushed in very quickly. The second feather will be moving towards the end of the hose. As it gets closer it will move faster until it is pushed into the hose. The 3rd feather will move slowly towards the hose. When it gets closer it will move faster and faster until it is pushed into the hose.

You can't see the air but you know it's there. You can't see the air go around and through the feathers but you know it's there. You can't see the air push on the feathers but you can see the effects of it. If you looked at the feathers from a distance, it would look like they are being pulled into the vacuum cleaner hose. Anyone can do this experiment. As we will see in the next couple of pages, this is how gravity works, except on a smaller scale.

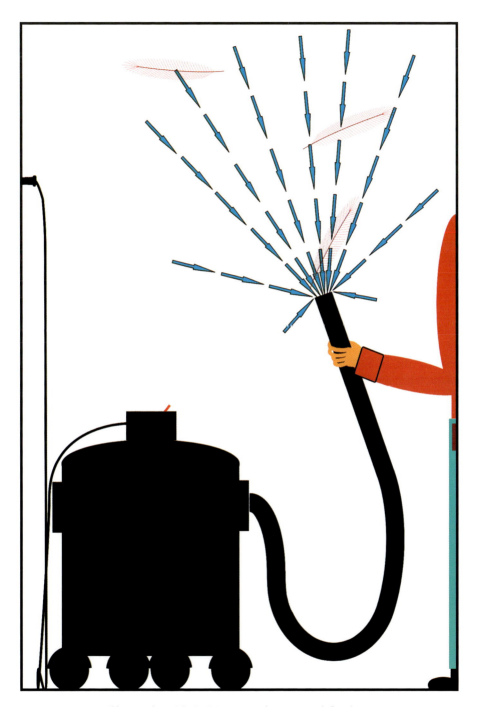

Illustration 13-2: Vacuum cleaner and feathers

Atoms being pushed by other atoms

On this page are the nuts and bolts of gravity. This is the core of how gravity works. We saw in Law 12 how a stream of gravitons can go through the center of all atoms. In the illustration on the next page we have 4 Hydrogen atoms. The Hydrogen atom at the bottom is our main emphasis. It has several streams of gravitons. The gravitons are the blue arrows. There are 3 other Hydrogen atoms above the main Hydrogen atom. They have the blue arrows going through them.

How does gravity work? It works just like the end of a vacuum cleaner hose. The gravity goes through the hole in the center of the atom. They go in the side of the atom that has the electron rings rotating towards the center. All available tadtrons turn to gravitons and follow in a stream. These streams now go through and around the other atoms. The gravitons push anything in their path.

How do the gravitons push? The gravitons are very thin. They can go through any matter. When they go through another atom, they bump into a proton ring or an electron ring. When they bump, a very slight push in the direction they are traveling is the result. If I throw a baseball at you and it hits you, you will be pushed in the same direction the baseball was thrown. It's the same as the feathers being pushed into the end of the vacuum cleaner.

It's hard to imagine that gravity is this simple but, it is. We will continue on and build gravity from the atomic level to the universal level. The question marks that are coming out the bottom will answer many things. The question marks are exiting gravity. We will cover that in Law 19.

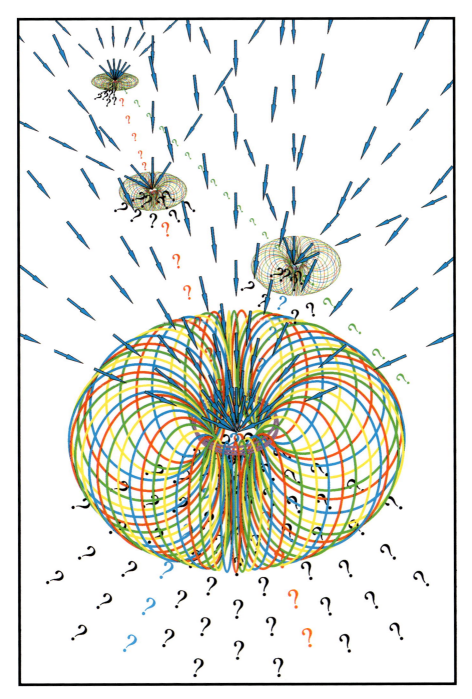

Illustration 13-3: Hydrogen Atoms

Earth and it's gravity

The next page is an illustration of the Earth, a plane, a space station and the moon. The blue arrows represent gravity. This is our next step up from the atom. What are the mechanics of gravity on the Earth? It works the same as at the atomic level. It works the same as the vacuum cleaner and the feathers. The difference is in the number of atoms.

How does gravity work on the earth level? Think of it as a huge number of vacuum cleaners. They are pulling in a huge amount of air. The Earth has a huge number of atoms. All the atoms have gravity streams. The combined number of gravity streams, produce the gravity on this Earth. When you walk on this Earth, billions of gravity streams may be going through you. They are all pushing you towards the center of the Earth.

The objects in the illustration are all pushed by the gravity of the Earth. The plane is the closest to the Earth. It has more gravity streams going through it. It has the most push by the gravity. The space station is much higher. It only has a few gravity streams going through it. Gravity has a much lower push on the space station than it does on the plane. The moon is farther out than the space station. It has even fewer gravity streams going through it. There are enough to hold it in orbit around the Earth.

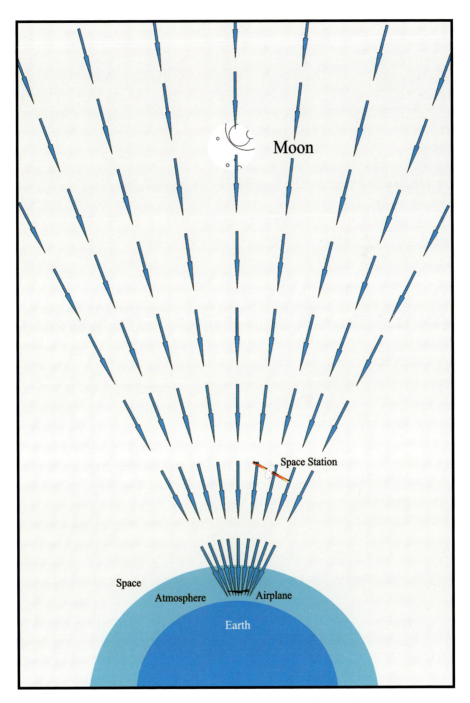

Illustration 13-4: Earth, plane, space station and moon

The Sun and it's gravity

The next page is an illustration of the Sun, Mercury, Venus and Mars. The blue arrows represent gravity. This is our next jump up from the Earth's gravity. How does gravity work on the Sun? It works the same as at the atomic level. It works the same as at the Earth level. One big difference. The Sun is bigger than the Earth. It will have many times more gravity streams.

If you could walk on the Sun you would weigh many times more than you do on the Earth. The reason would be due to more gravity streams that go through you. If you have more streams, you have more push effect. If the object being pushed is bigger, then the push effect is bigger.

The objects in the illustration are all pushed by the gravity of the Sun. Mercury is the closest to the Sun. It has more gravity streams going through it. It is pushed more by gravity. Venus is farther away than Mercury. It has fewer gravity streams going through it. Gravity has a much lower push on Venus than it does on Mercury. Earth is the third planet from the Sun. Gravity pushes less on the Earth than it does on Venus or Mercury. The Earth has fewer gravity streams pushing on it.

When something is in orbit around another object, it has a forward speed. If there is a higher push of gravity, it will allow for a higher forward speed. Objects closest to the Sun, orbit or go around the Sun much faster, than objects farther away. There is a relationship between forward speed, closeness of the objects and the size of the objects. The math on this has already been done.

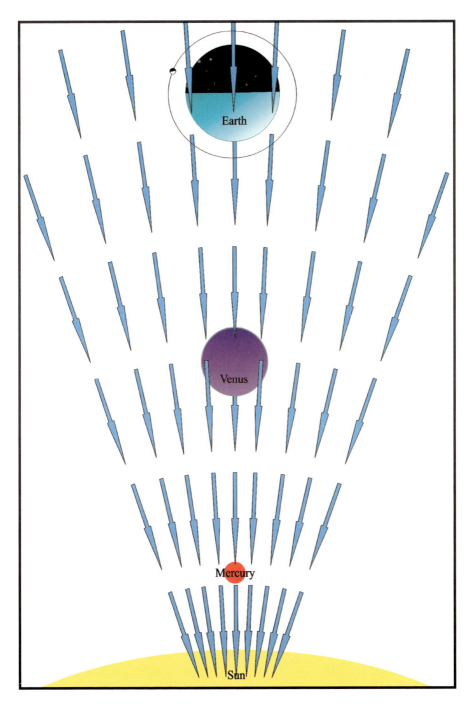

Illustration 13-5: Sun, Mercury, Venus and Earth

Law 13: Closing comments

This is a simple concept. Gravity pushes instead of pulling. Think of tornados. Their winds push and destroy things. The winds from hurricanes push and destroy everything in their path. There are no magical forces that pull. There are no illustrations of the mechanical workings of gravity pulling.

There are two concept we have not covered. One is what happens to the exiting gravity? We will cover this in Law 19. The second is what is the length of a gravity steam? We will cover that in Law 14.

Law 14

Law 14

A Gravity streams length can reach as far as light

Law 14: Opening comments

This is part 3 of the gravity section of the jigsaw puzzle. One of the big questions that hasn't been answered is, how far can gravity go? How far can gravity travel from one object and affect another object? These are the questions we are going to answer in this law. Don't put a limit on this. Open your mind.

How far can gravity from one object affect another? It will reach the same as the length of the gravity stream. How long are gravity streams? All atoms have gravity that pass through them. I will name this personal gravity. We all have personal gravity that travels through us. There is no way of knowing how long our gravity streams are. Let's look at objects that we know are affected by gravity.

The Earth's personal gravity affects a plane that is 1 mile above the Earth and a space station that is a few hundred miles above the Earth. The Earth's personal gravity affects the moon which is 238,855 miles above the Earth. The moon's personal gravity causes tides on the Earth. Gravity streams are at least hundreds of thousands of miles long.

The Sun's personal gravity affects the planets. Mercury is 36 million miles from the Sun. Venus is 7.2 million miles from the Sun. Earth is 92.9 million miles from the Sun. Pluto, the most distant planet, is 3.671 billion miles from the Sun. Comets go even farther. Gravity streams can be at least several billion miles long. Keep your mind open at this point. Think limitless!

Introducing the Galactic Super Mass

How far can gravity travel? How far can light travel? The answers are the same! Think limitless. We have amazing new telescopes. We are seeing things we have never seen before. There are an untold number of galaxies out in space. There may be hundreds or billions of galaxies. Each galaxy may have hundreds or billions of stars.

What is a galaxy? It's a group or cluster of stars. Have you ever noticed how they look? Most galaxies look like a large solar system. What holds the galaxies together? Gravity does. How? Just like the Sun and planets do in our solar system. That means there must be something at the center of each galaxy. What?

I was watching a program on TV. It was about a super massive black hole at the center of our galaxy. It had been identified as the center and as a theoretical black hole. The big problem was that light came from the black hole. If light is coming from it, then it isn't a black hole. I don't believe black holes exist, so it is something else. I will name this thing that is at the center of each galaxy a Galactic Super Mass.

The Sun is millions of times bigger than the Earth. The galactic super mass is millions of times bigger than the Sun. The illustration on the next page shows a galactic super mass and how it holds 3 solar systems in orbit. It works the same as the Sun and the planets. The only difference is it is bigger and the gravity reaches farther. The stars orbit the galactic super mass. This means that gravity streams can be at least several hundred light years long.

Several astronomers have already proposed this idea that solar systems are in a galactic orbit. Due to the lack of understanding of gravity these ideas have not taken off.

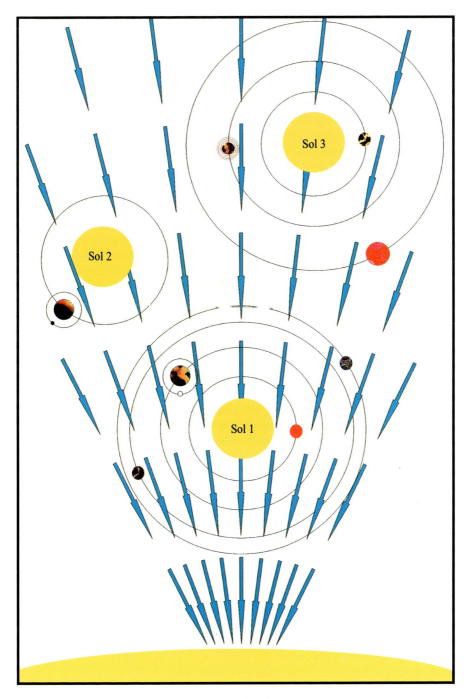

Illustration 14-1: Galactic Super Mass

Introducing the Universal Super Mass

Think about how far light can go. If you can think really big and really long, you can see this unifying theory. We will now unite the smallest molecule to the largest object in the universe. What is the largest object in the universe? It hasn't been discovered yet, but it will be. Nobody is looking for it. Nobody knows it exists, until now. What is it? Let's build up to it.

There is a big bang theory. Nothing exploded and created the universe. This is where the expanding universe idea came from. Some astronomers have evidence that the universe is expanding. Some galaxies are getting farther apart. Some astronomers have evidence that the universe is contracting or getting closer. Some galaxies are getting closer to each other. What is the contradiction?

If you where comparing the distance between Earth and Venus, what would you come up with? Different numbers every day. Why different numbers? Earth and Venus are in 2 different orbits. Sometimes Earth and Venus will be getting closer to each other. Sometimes Earth and Venus will be getting farther apart. If you thought our solar system was expanding, you would have evidence when Earth and Venus get farther apart. If you thought our solar system was contracting, you would have evidence when Earth and Venus get closer.

Is the universe expanding or contracting? No, it's an orbiting system. The universe is like our solar system. The universe is like the galaxies. The only difference is there is something at the center of the universe. It's at least a million times bigger than a Galactic Super Mass. It has gravity streams that reach for billions of light years. It's the Universal Super Mass. Each atom in the Universal Super Mass contributes to holding the universe together. The next page is an illustration of the Universal Super Mass and how gravity works. Gravity affects each Galactic Super Mass and holds each galaxy in orbit.

Illustration 14-2: Universal Super Mass

Law 14: Closing comments

Isn't this easy? Gravity works the same everywhere. The length of the gravity stream is the only limit to how far gravity can affect something. Gravity works the same from the smallest molecule to the largest object in the universe. The size and position of an object may affect the length of a gravity stream. If I move, my personal gravity streams will have to shift with me. This could cause the gravity streams to be shortened or changed in length. Objects that are in orbit may have shorter gravity streams than the objects that are stationary. The Universal Super Mass is probably the most stationary object in the universe. It probably has the longest gravity streams. The Galactic Super Masses are the next most stationary objects. They have the next longest gravity streams. This would continue on down from the stars, to the planets, to the moons.

Do the Galactic Super Masses exist? I have seen enough evidence to believe they do exist. What I have written may cause researchers to look for a Galactic Super Mass. Does a Universal Super Mass exist? In my theory it does. Everything else is in orbit around the Universal Super Mass. We should have enough older data on galactic postitions, to match with the current data, to find galactic orbits. If the orbits are established, then someone could find the center of the universe.

How fast is gravity? It may be slower than the speed of light. It may be millions of times faster than the speed of light. I believe gravity is faster than the speed of light.

What happens to the gravity after it gets to where it is going? What happens to the gravity after it passes through a molecule and becomes exiting gravity? Law 19 will discuss exiting gravity.

Law 15

Law 15

Sound is a particle in a corkscrew energy state

Law 15: Opening comments

Sound fills the air around us. Sound can be caused by vibrations. Sound can cause vibrations. The sound vibrations can go through solids, liquids, and gases. We are able to decipher these sounds when they reach our ears. We have eardrums in our ears. When sound reaches the eardrum, it causes a vibration. There are also tiny hair cells that are vibrated. The rest of our ear translates the vibration and the movement of the hair cells and sends a signal to our brain. Our brain then tells us what the sound is.

Sound causes the eardrum to vibrate back and forth. Sound causes the hair cells in your ear to be moved up and down and back and forth. The back and forth movement is believed to be caused by the frequency of the sound vibration. The up and down motion is believed to be caused by the pitch. Sound is thought to be like a wave in water. The distance between the waves is the frequency. The height of the waves is the pitch. There is one big problem with the wave idea. Sound runs at different speeds through solids, liquids, and gases. A vibration shouldn't change speed unless there was something behind the vibration.

In this chapter let's put a tadtron behind sound. This tadtron will be able to cause frequency, pitch, and change speeds in solids, liquids and gases. Read the next several pages and see how simple this is.

I am also going to introduce you to another sound that has not been identified. This sound works on the same principle as regular sound. This new sound causes vibrations that you can't hear, but you can see these vibrations.

How a propeller acts

Let's imagine you have a propeller that has no mass or momentum. The propeller only spins. It can only go forward when it catches something that has mass. The next illustration has four examples of propellers in different mediums.

In the first example, we have a propeller in a vacuum. The propeller will spin forever and go nowhere. It has nothing to propel it forward. Sound also will not travel through a vacuum.

In the second example, we have a propeller in the Earth's atmosphere. When the propeller spins it will go forward. It will be limited to how fast it will go because of how thick the air is. If you go higher in the atmosphere, the air becomes thinner and the forward motion will go slower. Sound travels better in thicker air.

In the third example, we have a propeller in water. When the propeller spins it will go forward much faster than in the air. There is more mass to push it forward. Sound travels about 2 times faster in water than it does in the air.

In the fourth example, we have a propeller in butter. You assume the propeller can move through the butter without friction. When the propeller spins it will move forward at its top speed. Each turn will have no give like water or air. Butter is a solid like aluminum. Sound can travel through aluminum 15 times faster than through the air.

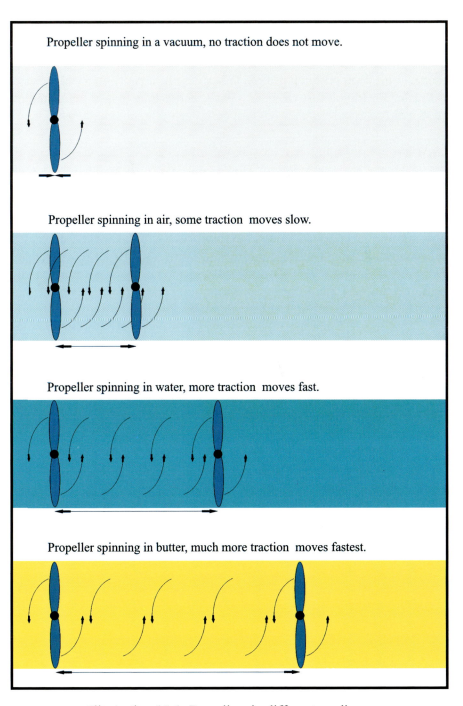

Illustration 15-1: Propellers in different mediums

How sound acts

Sound is a tadtron shaped as a corkscrew. It has no mass. It has no momentum. The corkscrew tadtron only spins. It can only go forward when it catches something that has mass. The next illustration has four examples of 4 tadtrons as corkscrews that propel through different mediums.

In the first example, we have a corkscrew tadtron in a vacuum. The corkscrew tadtron will spin forever and go nowhere. It has nothing to propel it forward. Sound will not travel through a vacuum.

In the second example, we have a corkscrew tadtron in the Earth's atmosphere. When the corkscrew tadtron spins it will go forward. It will be limited to how fast it will go because of how thick the air is. If you go higher in the atmosphere, the air becomes thinner and the corkscrew tadtron will go slower. Sound travels better in thicker air. The corkscrew tadtron will cause a vibration in the molecules as it travels through the air.

In the third example, we have a corkscrew tadtron in water. When the corkscrew tadtron spins it will go forward much faster than in the air. There is more mass to pull it forward. Sound travels 2 times faster in water as it does in the air. The corkscrew tadtron will cause a vibration in the molecules as it travel through the water.

In the fourth example, we have a corkscrew tadtron in aluminum. The corkscrew tadtron can screw its way between proton rings and electron rings. As the corkscrew tadtron spins it will move forward at its top speed. Each turn will have no give like water or air. Sound can travel through steel and aluminum 15 times faster than through the air. The corkscrew tadtron will cause a vibration in the molecules of aluminum as it passes through.

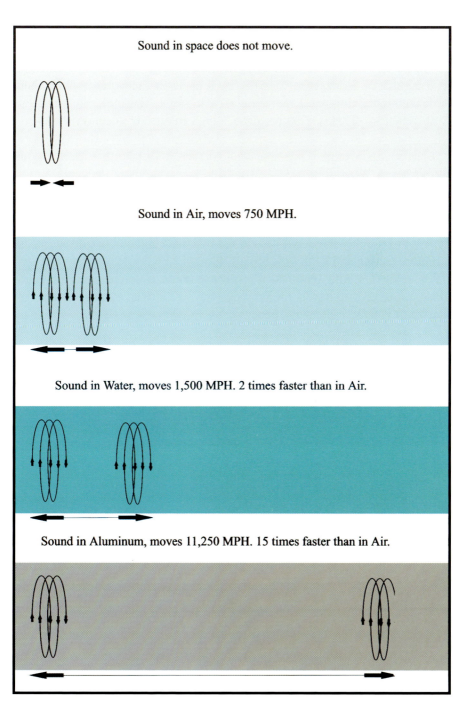

Illustration 15-2: Sound as a corkscrew

Frequency

There are two examples in this illustration. The first example is sound as a corkscrew tadtron that has the coils close to each other. When this corkscrew goes through a molecule, it will vibrate the molecule back and forth very fast. This is what produces a high frequency in sound.

The second example is sound as a corkscrew tadtron that has the coils farther apart. When this corkscrew tadtron goes through a molecule, it will vibrate the molecule back and forth slower than in example one. This is what produces a low frequency in sound.

It's that simple. Frequency is caused by how a corkscrew tadtron moves a molecule back and forth as it travels through. The corkscrew is very similar to a wave. Rather than waves, you have coils. If you have coils that are closer together, you have a higher frequency. If you have coils that are farther apart you have a lower frequency.

Illustration 15-3: Frequency

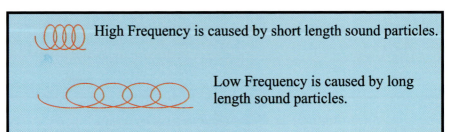

Pitch

There are two examples in this illustration. The first example is sound as a corkscrew that has the coils in a large loop. When this corkscrew goes through a molecule, it will vibrate the molecule slowly up and down. This is what produces a low pitch in sound.

The second example is sound as a corkscrew that the loops of the coils are smaller. When this corkscrew goes through a molecule, it will vibrate the molecule up and down faster than in example one. This is what produces a high pitch in sound.

It's that simple. Pitch is caused by how a tadtron moves a molecule up and down as it travels through. Once again this is very similar to a wave. The coils are now like the height of a wave. If you have coils that are taller, you have a lower pitch. If you have coils that are shorter you have higher pitch.

Illustration 15-4: Pitch

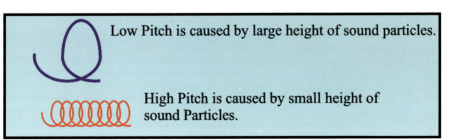

Volume or intensity

There are two examples in this illustration. The first example is sound as 4 different corkscrew tadtrons. This is a sound that would be quiet. When this sound reaches you ears, it is going to move your eardrum in and out. It is also going to move the hair cells up and down and in and out.

The second example is sound as 12 corkscrew tadtrons. This will produce a sound that is much louder than example one. The more corkscrew tadtrons you have, the louder the sound is. Loudness is also called intensity. What is going to happen when all these sounds go in your ear? The eardrum is going to move in and out more than example one. The hair cells are going to vibrate more. Your brain is going to tell you this is a louder sound.

Illustration 15-5: Volume or intensity

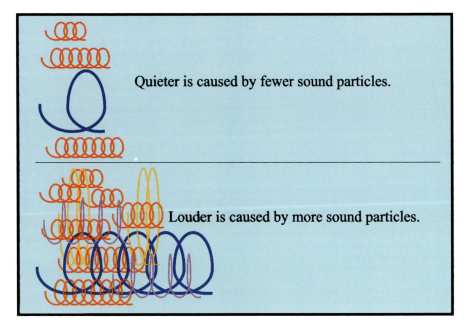

Ultrasound

There is ultrasound that humans can't hear. They have high frequencies and high pitches. Some animals, like dogs, can hear them. The ultrasound can be used in many applications. Ultrasound can be directed in a straight line. Ultrasound is reflected from objects it comes into contact with. Those reflections can be picked up by special instruments. Examples of these are sonar, fish finders, depth finders and medical ultrasound equipment.

The illustration we have is a boat with a fish finder. There are fish underneath. The fish finder is sending out sound that has a very high pitch and a very high frequency. The ultrasound tadtrons are a very tight corkscrew. The ultrasound tadtrons go easily through the water, but they are reflected from solid objects. The ultrasound tadtrons bounce off the fish and go back to the fish finder. The fish finder picks up these reflections and shows them on the screen.

Illustration 15-6: Fish finder

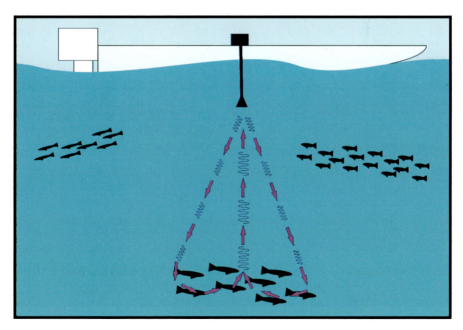

Molecular movement of sound

On the next page are three examples of sound at the molecular level. In each example we have sound as magenta in corkscrews. Watch as sound goes through each one. See how it will cause a vibration up and down and a vibration back and forth.

The first example is sound corkscrewing through an Oxygen molecule. Notice the gap in the middle of the Oxygen. The sound tadtron is not going to get a good grip as it corkscrews through. Like a propeller in the air, it will not be going at its top speed. The Oxygen molecule will vibrate as the sound tadtron corkscrews through it.

The second example is sound corkscrewing through a water molecule. Water is more dense than Oxygen. The sound tadtron is going to get a much better grip on the water than the Oxygen. What is the result? Like a propeller, sound will go faster in water than in air. The tadtron is going to cause a vibration in the water as it corkscrews through it.

The third example is sound corkscrewing through an Aluminum molecule. Aluminum is more dense than water or air. Is sound going to be able to go faster? Yes, the corkscrew tadtron is going to get a very solid grip as it travels through the Aluminum. The corkscrew tadtron will cause a vibration in the Aluminum as it passes through. A propeller can't travel through Aluminum, but if it could, it would travel much faster than air or water because of the additional density.

If sound goes through a solid, a liquid or a gas, why does it still sound the same when you hear it? It has changed speed several times, yet you still hear the same sound. Why? Because the sound tadtron is the same when it reaches your ears as when it was made.

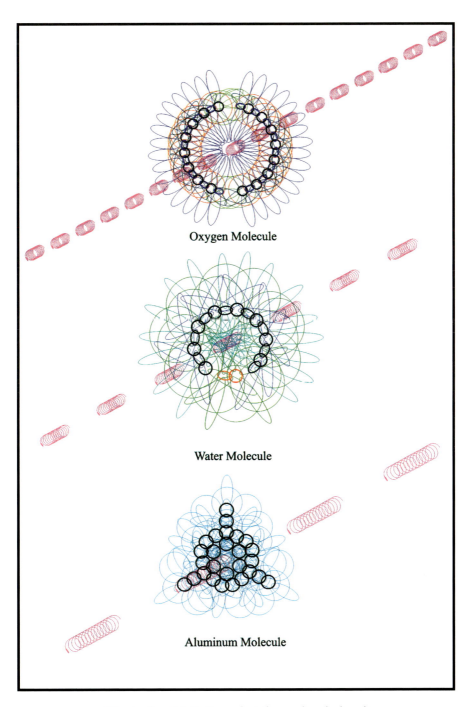

Illustration 15-7: Sound at the molecule level

Slow sound

There is another vibration that has not been identified as sound. It is on the opposite end of the scale from ultrasound. This sound will be much slower. Both the frequency and pitch will be very low. Like ultrasound, we are not able to hear these sounds. What did sound do before? It caused motion up and down. It caused motion back and forth. It ran as a corkscrew. So, where do we find this new sound? It's in the waves of the water. Watch some waves. They go up and down and back and forth. I will call this new sound, "Slow Sound".

On the next page are three illustrations of waves, with sound in them. Illustration one is a side view of water waves with tadtrons. It looks like a regular wave. The second illustration is a 45 % angle view of water waves. See how the tadtrons are corkscrewing through the water and causing the waves. The tadtrons are like a school of fish causing the waves in the water to rise and fall. The third illustration is a look at the waves from the back. You can see how slow sound goes in a corkscrew causing the waves.

It's that simple. Slow sound is the energy in the waves. The waves can be small or large. Slow sound will continue to travel in the waves of the water until it comes in contact with something that stops the sound. If you were out in a boat and there were hundreds of waves, and if you could tell all the slow sound tadtrons to change their state, the water would instantly become calm.

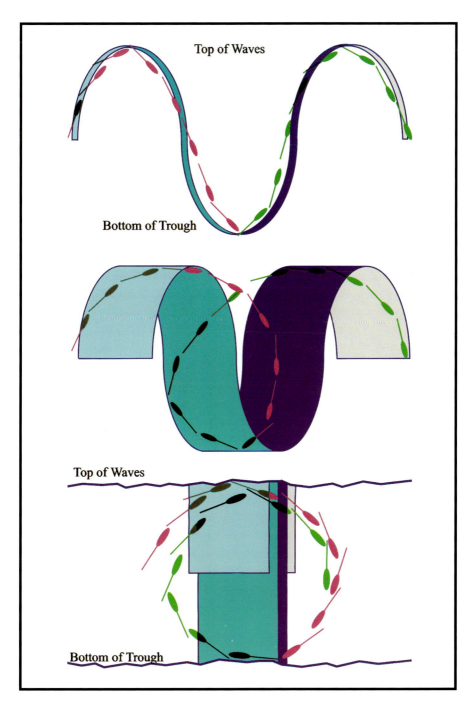

Illustration 15-8: Waves with slow sound

Boat produces slow sound in the waves

What causes slow sound? This is not a difficult question to answer. Slow sound is caused by a couple of things. One thing is the wind. If you have a wind, waves will start to move in the same direction the wind is blowing. The longer the wind is blowing the bigger the waves will get. A stronger wind will produce bigger waves. The waves come from objects colliding or coming in contact with a liquid. The water is the liquid. The wind is a gas. When the wind collides with the water, slow sound is the result.

Another thing that causes slow sound is a solid object moving in a liquid. In the illustration on the next page, we have an aluminum boat traveling in a body of water. When the boat moves through the water, waves are produced. The size and shape of the wave are a result of the size, shape and speed of the boat. A bigger boat and faster speeds produce bigger waves. If the boat in the illustration is traveling at a constant speed in a smooth body of water, it will produce exactly the same waves wherever it goes. The waves are known as a wake. Along the side of the boat, I have tried to show the corkscrew patterns that produced the waves. Those corkscrews are the slow sound in the waves.

If you throw a rock in a calm body of water, waves will go out in all directions from where the rock hits the water. This is slow sound as well. There are particles in the slow sound. Where do they come from? We will answer that in Law 19.

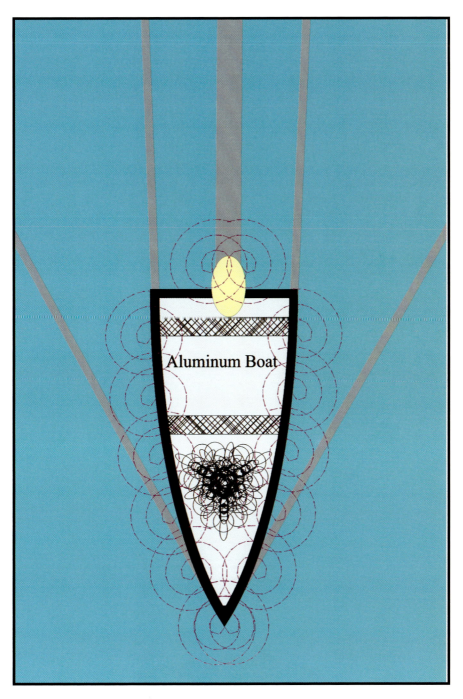

Illustration 15-9: Boat producing waves

Car honking horn causes Doppler effect

There is something called the Doppler effect. When a car comes toward you and is honking its horn, it will sound different than if the car is going away from you. In the illustration below we have this example, a car is driving along and honking its horn. There is a runner in front and a runner in back. Why does the honk sound different between the two runners? When an object is moving towards something and produces a sound, the tadtrons will be closer together. When an object is moving away from something and produces a sound, the tadtrons will be farther apart. The distance between the tadtrons will cause a difference in the sound you hear.

Illustration 15-10: Car honking horn causes Doppler effect

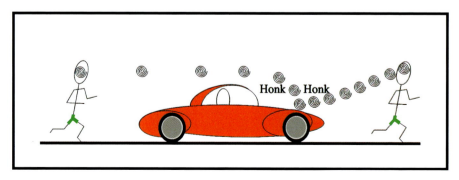

Law 15: Closing comments

One of the things we have not covered is sound being reflected. It's not difficult to image the tadtron corkscrew hitting something and bouncing off. It's like throwing a ball against the wall, it will reflect at an angle. Sound will be no different. Sound is sometimes absorbed, changing its state, and is no longer sound.

One big thing we have not covered is, where does sound come from? Where do the corkscrewing tadtrons come from? Law 19 will give this answer.

This is one more piece of the puzzle turned up. You should be able to see how sound can travel through solids, liquids and gases and yet still be the same sound when it hits your ears. When you see waves in water, imagine the little tadtrons corkscrewing to cause the up and down and back and forth motion. You can now see how waves, from different directions, can go through each other and then continue on.

Law 16

Law 16

Light is a tadtron in five different states

Law 16: Opening comments

Light is what gives us an ability to see. Sight may be the greatest of the five senses. If we don't have light, we can't see. Light must have a source. It must come from somewhere. The sun is the greatest source of light on the earth. Light comes from the sun and is reflected from many objects. We see the light that is reflected. We can see the color, size and shape of the object in the reflected light. This light is known as visible light. Light from the sun also causes heat. This light is known as infrared light. Another type of known light is ultraviolet light. Ultraviolet light causes sunburns.

There has been a big debate of what light is. One of the mainstream beliefs was that light is energy that has different wave lengths. Different wave lengths determined whether the light was ultraviolet, infrared or visible light. Different wave lengths also caused all the different colors in the visible light. Another belief was that light was packets of photon particles.

In this chapter I am going to give you tadtrons in 5 different states. How the tadtrons act in each state, will give easy examples to follow of what light is and how it works. We have the framework of our jigsaw puzzle. We have the gravity section most of the way together. The sound section is together. This is the light section. It is an easy section. Let's move on and assemble this section of the puzzle.

Light tadtrons

The illustration below contains tadtrons in 5 different states. The tadtrons are infrared light, red visible light, green visible light, blue visible light and ultraviolet light. Each of the tadtrons has a different shape on its head. The tails are all straight. The shape of the heads will be what give each tadtron its characteristics. These shapes are important and will be what identifies the particle. When tadtrons come in contact with the matter, the shape will determine how each particle is reflected or changed.

Illustration 16-1: Tadtrons in the light state

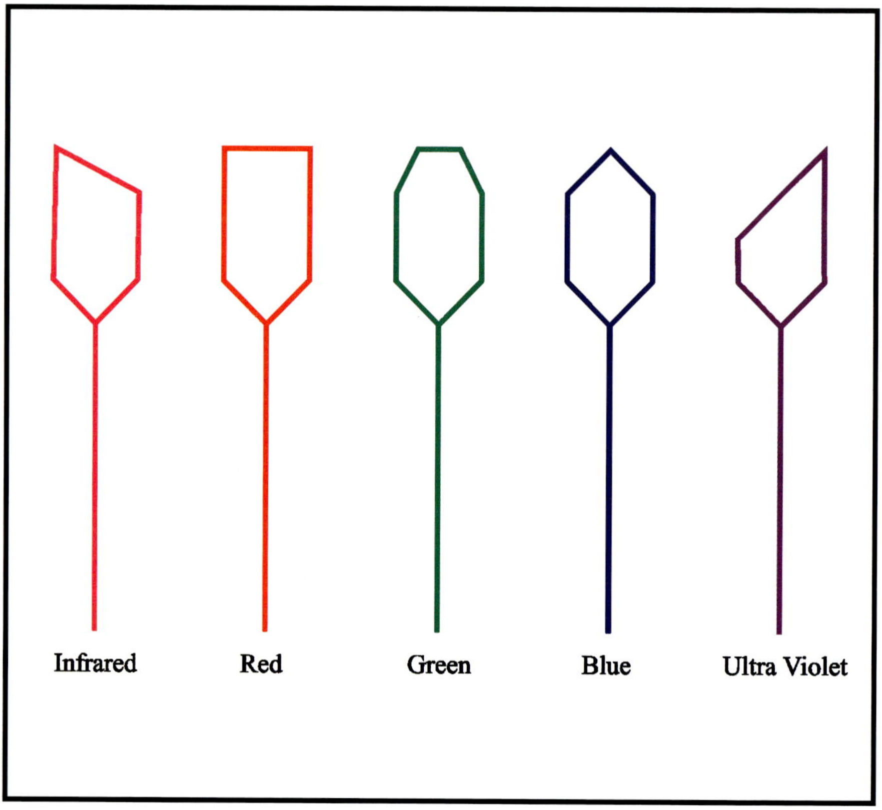

| Infrared | Red | Green | Blue | Ultra Violet |

Red, green and blue circles on a TV

Below is an illustration of 3 circles with overlapping colors. The top circle is green. The bottom left circle is red. The bottom right circle is blue. This illustration is emulating colors that come from a computer screen or a TV. TVs have only red, green and blue guns. These are the only colors they produce. In the yellow section your eyes are seeing both red and green. In the magenta section your eyes are seeing red and blue. In the cyan section your eyes are seeing both green and blue. In the white section your eyes are seeing red, green and blue. By mixing any combination of red, green and blue light, a TV or computer screen can reproduce almost any picture of any object.

Illustration 16-2: Red, green and blue circle

Light going through glass (a square hole)

On the next page is an illustration of light coming from a white light source, passing through glass and into an eye.

How does light work? First it must have a source. The object at the top could be a sun or a flashlight. It produces the red, green and blue light tadtrons. The tadtrons travel from their source until they come to the glass. Light is able to travel through the square hole in the glass. Things that light can travel through are called transparent. The light travels through the glass and into the eye. The eye can decipher the colors. In this instance, your eye would see white because it would pick up red, green and blue light tadtrons.

How does your eye work? Your eyes have rod and cone shaped cells. These cells pick up the different colors and send signals to your brain. Your brain then blends the red, green, and blue and you have a picture in your mind. This is how you can see. If someone is color blind, they cannot tell the difference between red and green. Their eyes see the red and green tadtron as the same. There are also some cells in your eyes that pick up on the intensity of the light. If the light is brighter, a stronger signal is sent to your brain. A whiter color is the result. When the light is weaker, everything appears darker.

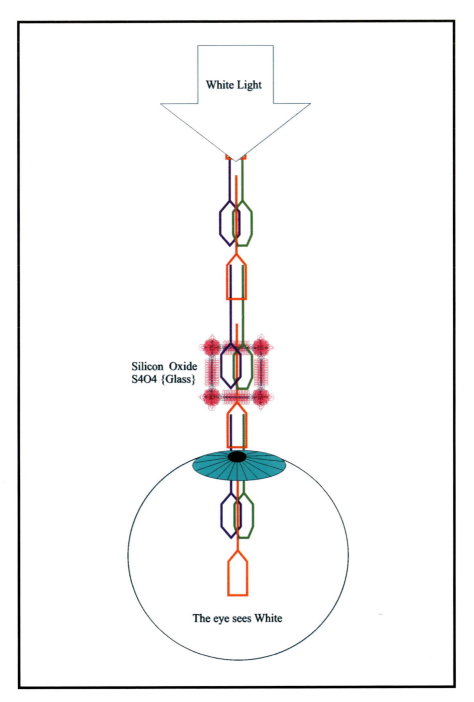

Illustration 16-3: Light, glass and an eye

Light going through a prism

In the illustration on the next page we have white light going through a prism. The white light is coming from the bottom. The prism is at the bottom of the page as well. At the top is a breakout of the 3 colors. There is a green bar, a red bar and 2 small blue bars. The color spectrum is at the top. A prism is a piece of glass shaped in a triangle. A prism causes white light to separate into a color spectrum. A rainbow is a result of a prism, made of water mist in the air.

The color spectrum is broken down one level further. Here there are 3 colored bars. When the white light passes through the prism it has equal numbers of red, green and blue light tadtrons. The light tadtrons are traveling in a straight line into the prism. When the red light tadtron comes out of the prism, it will be deflected mostly to the left. The deflection is due to the shape of the head. The red light tadtrons will form the red bar. The green light tadtrons will be deflected mostly to the right, due to the shape of their head. They will form the green bar. The blue light tadtrons will go to the far right and the far left. They will form the 2 blue bars.

The color spectrum behind the 4 bars is the result of a combination of the colors. Red, green and blue form all the colors of the color spectrum. The center of the red and green bars will have the most tadtrons of that color. The edges will have fewer. The blue tadtrons will have the highest concentration at the far left and the far right. All the colors are the result of combinations of red, green and blue tadtrons and their concentration in each area. It's that simple.

The lights that are not visible are deflected to both sides. Infrared light goes to the red side. The ultraviolet goes to the opposite side where there is more blue. The shapes of the tadtron heads will determine where they deflect.

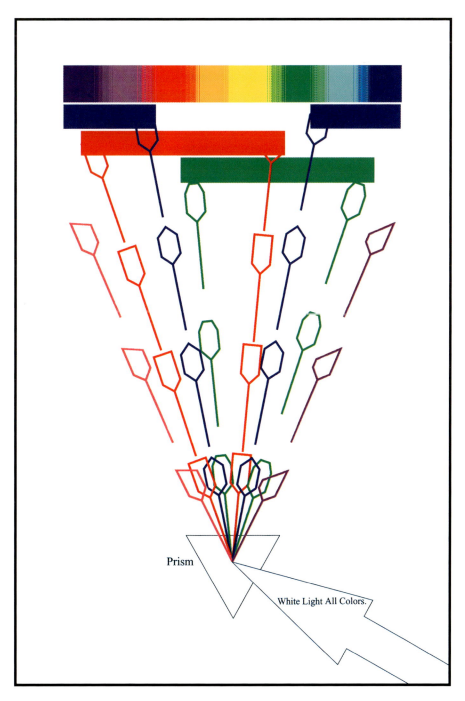

Illustration 16-4: Color breakout and color spectrum

Color reflected from leaves

The illustration on the next page is light being reflected from 3 leaves. The light at the top is white light. The middle has 3 leaves that are reflecting light. The bottom is an eye that receives the reflected light.

The leaf on the left is a green leaf. Three particles of light hit the green leaf. The red and blue light tadtrons change to an unknown state. They are no longer light. The green light tadtrons are reflected and go to the eye. The eye now sees it is a green leaf.

Leaves die and change color in the Fall. The leaf on the right is a yellow leaf. As a leaf dies, it gains the ability to reflect red light. The yellow leaf now reflects red and green. When red and green are both present, you will see yellow, unless you are color blind. The blue light still changes to an unknown state.

As the leaf continues to die, it loses its ability to reflect green light. The leaf in the center is the red leaf. It only reflects the red light tadtrons. Your eyes will see a red leaf. The blue and green light tadtrons are changed to an unknown state.

We will consider the unknown energy states in Law 20. They are still a tadtron. We just can't see them because they are not light anymore.

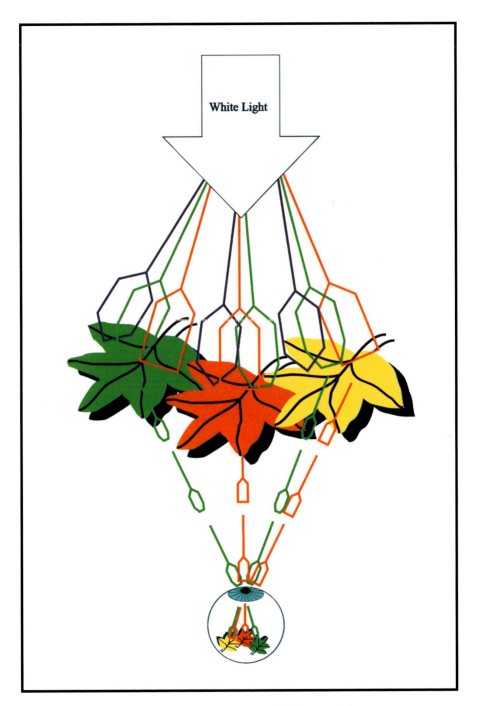

Illustration 16-5: Leaves and light particles

Light reflecting from 3 shirts

On the next page is an illustration of light being reflected from 3 shirts. White light is coming from the top. The 3 shirts are in the middle. An eye receives the reflected light at the bottom.

The first shirt on the left is a blue shirt. It only reflects blue light tadtrons. The green and red light tadtrons are changed. The shirt in the middle is a red shirt. It reflects red light tadtrons and changes the green and blue light tadtrons. The shirt on the right is a magenta shirt. It reflects blue and red light tadtrons. The green tadtrons are changed. The eye takes in the reflected light. It sends signals to the brain of the color of the shirts.

Light is that simple. The tadtrons are reflected, to cause the color. If an object doesn't reflect any light, what color is it? It's black. If an object reflects all the colors, what color is it? It's white. The white paper you see is reflecting all the red, blue and green light tadtrons. The black letters are reflecting no tadtrons. Your eyes register it as black.

When red, green or blue light tadtrons come in contact with an object they seem to have no effect on the object. They don't push against the object to cause it to move or heat up. The red, green and blue tadtrons simply bounce off, are absorbed, or changed to another state.

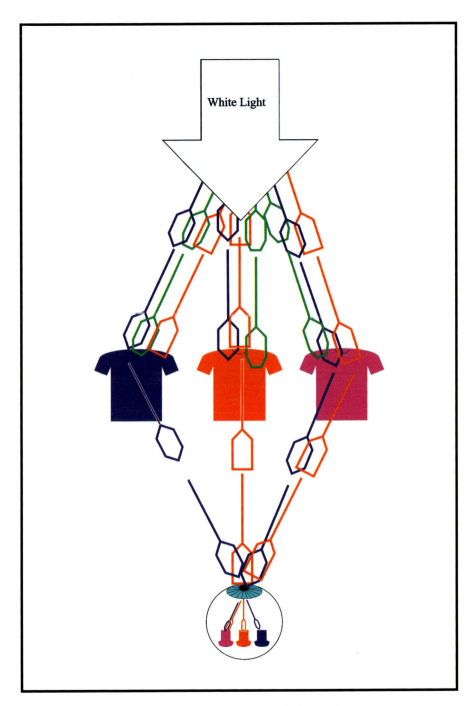

Illustration 16-6: T-shirts & light particles

Ultaviolet light and infrared light

The illustration on the next page has the sun at the top. The sun is producing ultraviolet tadtrons and infrared tadtrons. The ultraviolet tadtrons and infrared tadtrons are coming down to a sunbather on the beach by an umbrella. The ultraviolet and infrared tadtrons are coming into contact with his skin.

We can't see ultraviolet light. We know it's there. It was discovered many years ago. How does ultraviolet light work? It's a single particle. It has a head and a tail. It travels forward in a straight line. What happens when your skin is exposed to ultraviolet rays? You can get tanned or sunburned. Ultraviolet rays also help in the production of Vitamin D. You can put suntan lotion on and it will block the ultraviolet rays, so you won't get sunburned.

The infrared light is also coming in contact with your skin. What happens? Your skin will get hotter. The suntan lotion won't help to keep you cool. If you walk out of the sun and into the shade, how do you feel? You will feel cooler. When the sun goes down at night, what happens? Everything gets cooler. What happens when the sun comes up in the morning? Everything gets hotter.

How does infrared light work? It's a single particle. It has a head and a tail. It travels forward in a straight line. When infrared light strikes an object, it causes the object to heat up. How does it do that? It will cause the electron rings of the molecule it strikes to spin in a larger orbit. Infrared light is how heat is transferred from 1 object to the next, when the 2 objects are not touching. If you stand several feet back from a fire, you can feel the heat from the fire. The infrared light tadtrons are what you feel.

Illustration 16-7: Sunbather on the beach

Infrared light can be seen

The illustration below is at night in Uncle Mike's garden. There are rabbits in the garden. We are looking through infrared goggles. The infrared goggles are picking up infrared light. The rabbits are hotter than anything else in the garden. Their outline and their bodies are detected by the goggles. The goggles convert the infrared light to visible light. Uncle Mike isn't happy with the rabbits in his garden. He is firing a warning shot over the rabbits' heads. The blast from the gun is hotter than the rabbits so it shows up brighter.

Illustration 16-8: Infrared light at night

Law 16: Closing comments

One thing I have not covered is how the visible light tadtrons reflect. Visible light tadtrons can be traveling at the speed of light. They move forward until they hit a mirror. The tadtrons are reflected backwards. They can do a complete 180 degree turn. There is no damage to the mirror. The visible light tadtrons seem unaffected other than the direction they are going. The ability to have a speed that comes in contact with an object, reflecting back at the same speed without causing damage, is something I named Super Reflectivity. I am going to talk more about Super Reflectivity in Practical Applications.

What is the speed of the 5 particles? We don't know the exact speeds of all the particles. The speed of light is 186,000 miles per second. When light comes out of a flashlight, it is dispersed. If you shine a flashlight against the wall, most of the light will go straight forward. Smaller amounts of light will cover the whole wall. Something causes the light to break apart and go off in different angles. These particles may have slightly different speed. For example a red light tadtron may be 1 mile per second faster than a green light tadtron. A green light tadtron may be 1 mile per second faster than a blue light tadtron. What would be the effect of different speeds? It would be just like a highway full of cars. When a faster car from behind hits a slower car in front, what happens? They both break off at different angles.

Scientists have made lasers. Lasers are a concentration of 1 color of light. Lasers are able to shine forward without the dispersion. All the light tadtrons are going the same speed. They don't overrun each other.

What is the speed of the visible light tadtrons? If you are in a car going 70 mph, then you shine a light to the front from the back, those visible light tadtrons are traveling the speed of light. That is, if you measure the speed of light inside the car. What is the ground speed of those visible light tadtrons in the car? It's the speed of light plus the speed of the car. It's all in how the speed is measured.

What about paints and pigments? They work different than a color TV. Pigments have 3 primary colors. Those colors are red, yellow and blue. I will give a brief description of how they work. When you see yellow, what colors do you see? You see green and red reflected. Your eyes blend the colors so you see yellow. If you mix blue and yellow pigments, what do you get? You get green. Why? The blue pigment blocks the red pigment and all that is reflected is the green light tadtrons. The red and blue light tadtrons are changed. This is the principle of mixing paints and pigments.

This is one more section of the jigsaw puzzle. Light gives you the ability to see. It may seem a little odd, but combinations of 3 particles give you your greatest sense.

Law 17

Law 17

Electricity is a particle that swims

Law 17: Opening comments

Electricity is an important section of the jigsaw puzzle. Most of this section of the puzzle has already been put together. There are two pieces out of place. We are going to put them in their proper place.

Electricity is the one energy that has properly been identified as a stand-alone particle. There has been a large body of work compiled on electricity. In the past, the electricity particle was known as an electron. Most of the work on electricity is correct. A part that I disagree with is how electricity moves through matter. It was believed that the electrons jumped from atom to atom. They did this by sharing the orbit of the protons and neutrons. I don't believe in the proton and neutron, so I don't believe in this orbit.

I believe electricity is a tadtron in the electrical state. It has a head and a tail and it swims. I believe it looks very similar to a tadpole. I have no problem calling it an electron. In the next several pages, I am going to discuss conductors, insulators, voltage, amperes and circuits. Then I am going to show you how I think it swims through matter.

A battery in a flashlight

The illustration on the next page has 2 flashlights. The top one is turned off, the bottom one is turned on. The blue part of the flashlight is the batteries. Batteries are able to store and produce electrons.

We need to talk about some terms. The first term is a conductor. A conductor is matter that electricity can swim through. In the flashlights we have copper wires coming from the posts of the battery. There is a red switch at the top of each flashlight. When you press the switch, it connects the copper wires. This makes a circuit. A circuit is nothing more than a connection of conductors. Once the conductors are connected, the tadtrons can now swim through the conductor.

Now, two more terms. One is voltage, which is the amount of pressure on the tadtrons. The second term is amperage, which is the measure of the number of tadtrons. Another way to say this is amperage is the number of tadtrons available to do something. Voltage and amperage are controlled in many ways. The control of voltage and amperage is well known in the electrical industry.

When a flashlight is turned on, the circuit is made. The voltage of the battery causes the tadtrons to swim into the copper wire. They swim from the plus to the minus. Some of the tadtrons going through the light bulb are changed to light tadtrons, and leave the circuit. The amount of stored amperage, in the battery, determines how long the flashlight will last. The top flashlight has fewer tadtrons in its battery than the bottom one. One way this is measured is ampere hours. The bottom flashlight has the most ampere hours.

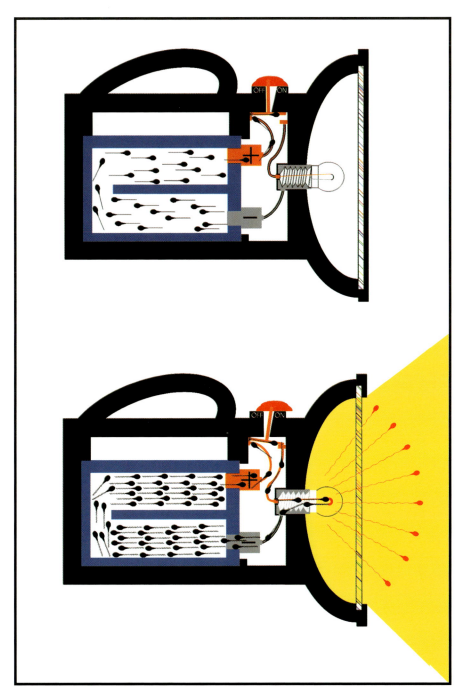

Illustration 17-1: Flashlights "off and on"

Solid and liquid conductors and insulators

There are 4 things in the illustration on the next page. At the top is a copper wire coated with plastic. Below it, is a plastic pipe with water in it. On the bottom left is a copper wire in a pitcher, pouring tap water on a table. On the bottom right is a copper wire going into a glass of distilled water that is sitting on a table.

I need to give you a definition of an insulator. An insulator is matter that will not let an electron swim through it. The copper wire at the top is a conductor, electricity is swimming through it. The plastic coating is an insulator. The electron can't get out of the copper wire and go through the plastic.

The plastic pipe has tap water in it, the water is the conductor. Electrons are swimming in the tap water. The plastic pipe is the insulator. The electrons can't swim through the plastic pipe.

At the bottom left, the copper wire and the tap water are conductors. The electrons can swim through the copper wire and through the tap water. A circuit is completed when it grounds to the table. The water pouring out of the pitcher completes the circuit. The glass pitcher is an insulator and the electrons can't swim out except in the stream of water, which is the conductor.

At the bottom right, the distilled water is an insulator. The electrons from the copper wire will not swim through the distilled water. I have shown 2 solids as insulators, glass and plastic, and 1 liquid insulator, distilled water. I have also shown a solid conductor, copper, and a liquid conductor, tap water.

I suggest you do not try to do the distilled water test. If there is any contamination, the water will conduct electriciy.

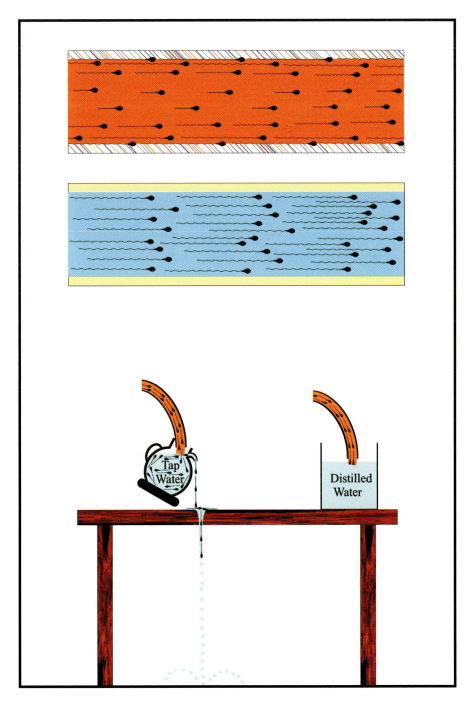

Illustration 17-2: Copper wire, pipe and pitchers

A gas can be a conductor or insulator

The illustration on the next page is a lightning bolt coming out of the clouds or through the air. There are some combinations of gases that are able to become a circuit. Lightning forms a circuit from the clouds to the ground. No one really knows if the electricity is going up or down. The illustration shows the electrons going both ways.

Lightning has a massive amount of voltage and amperage. Lightning is extremely powerful and dangerous. It has never been harnessed. Where the electrons come from, has never been fully understood. With my theory, there are some unseen tadtrons that convert to electrons. These electrons are able to form a conductor in the air, generating extreme voltage, and the result is a bright flash of light. Many of the electrons get converted to light tadtrons when lightning strikes.

The main point I am making is that a gas can be a conductor. Also, a gas can be an insulator. If you have a bare wire with current running through it, the electricity usually does not go out of the wire. Under some circumstances a bare wire with current running to it can arc to a ground. In this circumstance, they are similar to lightning, except on a smaller scale.

A gas can be a conductor or an insulator. Neon gas is a good example of a gas being a conductor and part of a circuit.

Illustration 17-3: Lightning

Electricity at the atomic level

The illustration on the next page has 4 molecules of Copper. The illustration is showing how it is possible for electricity to travel through a solid. The Copper has black proton rings and red electron rings. The electrons are black tadpoles.

As you can see, the electrons can swim through and around the outside of the Copper. There is a hole in between the Copper molecules. The electrons can swim in this hole. The hole is in between 4 round Copper molecules. In a sense it is a round hole. Electricity can go through a round hole. The ability to go through this hole makes Copper a conductor. Electricity is also able to swim on the outside edges of the Copper.

The medium that electrons can swim through is the outside edge of an atom. The outside electron rings are able to hold the electrons in place. The electrons have a head and a tail. They can swim from 1 set of electron rings to the next. If a Copper molecule is touching another Copper molecule, the electrons can swim in between them. This is what a conductor is. It does not matter whether it is a solid, a liquid or a gas. If there are some electron rings electricity can swim in, then it is a conductor.

I have shown some glass in previous illustrations. The glass had square holes in it. Electricity won't swim through the square hole. If any molecules have an arrangement of electron rings that electricity can't swim through then that material is an insulator.

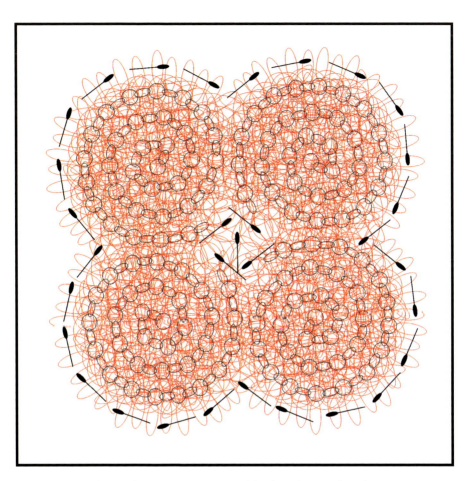

Illustration 17-4: Copper with electricty swimming

Law 17: Closing comments

The way I have presented electricity is an easy concept. Most of what is shown is with direct current. Direct current goes one way. It's interesting that they call the movement of electricity a current. It is like a flowing river. The second kind of current is called alternating current. With alternating current the electrons go back and forth many times a second. Alternating current will travel much farther in a conductor than direct current.

There are thousands of books on how electricity works. Electricity is our most usable form of energy. We have harnessed it. We can control it. We can transport it. We move things with it. We run our lives with it. We need it. I hope I have helped you understand electricity a little better. If we can find better ways of converting other energies to electricity, we could improve the lives of all on this planet.

Law 18

Law 18

Magnetic fields are caused by circling energy particles

Law 18: Opening comments

Magnets have been around for thousands of years. The earliest record of magnets is around 500 BC. Greek literature records magnets or lodestones as having the ability to attract and repel. The first recorded use of magnets as a compass is from around 1000 AD. The compass was used in navigation.

Magnets have always puzzled me. They can push apart. They can pull together. There are supposedly forces that cause this phenomenon. I have never seen an adequate answer for how magnets work. I started this theory trying to understand how magnets work. Strangely enough, this is one of the last things I found an answer for. The answer is not complicated when you apply some of the previous ideas in this theory. Let's go on and see how magnets really work.

I am going to give you a little heads up on what is behind the scenes of a magnet. I want you to think of a common configuration of animals. Most animals have a skeleton. In the skeleton there is a spine. The spine consists of individual circular pieces of bone. The circular pieces of bone are known as vertebra. Anchored to the spine is a series of ribs. The ribs circle out from the spine. Each rib circles around and attaches to a single vertebra. This configuration is very durable and stable. The spine can be bent in several directions. Some people are even flexible enough that they can bend into the shape of a horseshoe.

Electromagnets

In the early 1800's several scientists produced magnetism with electricity. This was called an electromagnet. Scientists made electromagnets by wrapping wires around an iron needle and then applying an electrical current. The electromagnets could lift many times their own weight. Stronger magnets could be made by wrapping more coils around the needle.

The next illustrations show an electromagnet. Anyone can build an electromagnet. All you have to do is get some coated copper wire and wrap it around an iron nail. Then hook each end of the wire to a standard direct current battery. You will have an electromagnet. Take away the current and it will stop working. If you add more wraps of wire, the electromagnet will be stronger. If you use a stronger battery, you will have a stronger electromagnet. The iron nail can be straight or it can be bent. The electricity travels through the copper wire. A magnetic field now surrounds the nail. What is this magnetic field?

Look at the next 2 illustrations. Think of the copper wire as a spine. Electricity is circling through the copper wire. Electricity is a tadtron with a head and a long tail. The circling tadtrons will be used as anchor points for the ribs. I am going to name the individual circling tadtrons the Inner Magnetic Rings. The combination of inner magnetic rings I will name the Magnetic Spine. The magnetic field is caused by what I will now name the Magnetic Ribs. I am going to show you these magnetic ribs and how they work.

The reason I am showing the electromagnets is to help you visualize what is happening on the inside of a magnet. There is a spine of circling electricity. The magnetic field is formed around this spine. You cannot see the spine or the magnetic fields because light does not reflect from electricity or magnetic fields. In an electromagnet you can measure the electricity that goes through the copper wire. The north and south poles are at opposite ends of the magnet spines.

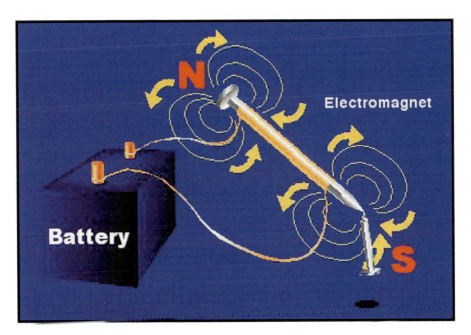

Illustration 18-1: Electromagnet with a nail

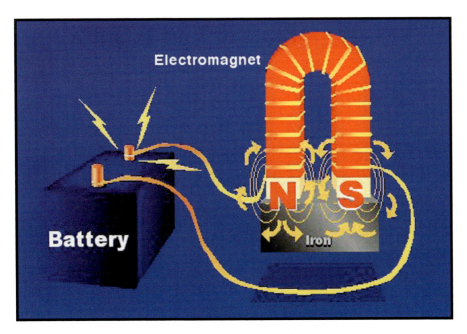

Illustration 18-2: Electromagnet with a horseshoe

Permanent magnets

We have shown an electromagnet. Now we are going to show permanent magnets. A permanent magnet always has a magnetic field. It does not need a source of electricity to make it work. The older magnets were iron ferrite. There are some neat new things with magnets. First there are some new magnets that are more powerful than the older iron ferrite magnets. These new magnets are known as rare earth magnets. They are made with Neodymium, Iron and Boron. They are very powerful. If you are working with big Neodymium Iron Boron magnets you need to be very careful. They will snap together and pinch or crush your fingers.

In the next 2 illustrations I have some Neodymium Iron Boron magnets. The first illustration is a picture of 2 Neodymium Iron Boron magnets. They are both 1 inch cubes. They are accidentally stuck together. I have been unable to separate them. The north pole on the magnet is facing to the left. The south pole is on the right.

The second illustration is a picture of a Neodymium Iron Boron washer magnet. The north and south poles are on the top and bottom edge of the washer. The poles are not like heads or tails of a penny. The north and south poles go through the circumference of the washer. The north and south poles are on the round part of the washer.

I picked these 2 types of magnets so I could show that the magnetic spine and the magnetic ribs are the same even if the magnets are different shapes.

The next neat thing is Ferrofluid. Ferrofluid is made up of very tiny round magnets, suspended in a black liquid. When a magnet gets close to Ferrofluid you can see an outline of the magnet fields. That is where we will be going next.

Illustration 18-3: 2 Neodymium Iron Boron 1 inch cube magnets

Illustration 18-4: Neodymium Iron Boron washer magnet

Ferrofluid in a magnetic field

The next illustration is a picture of ferrofluid in a magnetic field. There is about 50cc of ferrofluid in a plastic container. The magnets in this illustration are the two 1 inch cube Neodymium Iron Boron magnets. The north pole is directly beneath the plastic container.

Notice the cool looking ball that has formed in the ferrofluid. This is the magnetic field that is coming out of the north pole of the magnet. Look very closely at the ball of ferrofluid   You can clearly see little nubs or small columns of ferrofluid. The ferrofluid is trying to follow the magnetic ribs. You can see curves in the ferrofluid. The columns are trying to follow or form a circle. These circles that the ferrofluid try to follow have different circumferences.

If you take the magnet and pull it away from the bottom of the plastic container, the shape of the ferrofluid will change. The columns of ferrofluid continue to follow the circles or the magnetic ribs. The diameter of some of these circles or magnetic ribs are less than 1 inch. The diameter of some of these circles or magnetic ribs is over several feet.

Notice the diameter of the ball of ferrofluid. It's larger than the 1 inch face of the magnet. It is not square like the magnet. What makes the ball of ferrofluid? We will explain some of it in the next illustration.

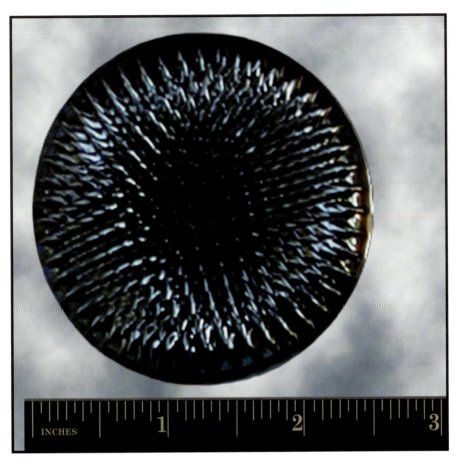

Illustration 18-5: The ball of ferrofluid

Inner workings of a magnet

Here we are going to show you the inner workings of a magnet. This illustration is a 2D image of an inner magnetic ring and 2 sets of magnetic ribs. This is a cross section of the inner workings of a magnet. The inner magnetic ring is a series of tadtrons in that energy state. The tadtrons have the job of circling inside of the magnet. That's what they do. Most of the magnets, that we have, are charged by a much larger electromagnet. Once the inner magnetic rings are formed, they stay. The Neodymium Iron Boron provides a medium by which this inner magnetic ring is stable. If you heat a magnet to a high temperature, the medium becomes unstable and the magnet will lose its magnetism. If a magnet receives a hard blow, it can lose its magnetism. So what is the inner magnetic ring? A series of circling tadtrons.

Circling tadtrons may seem like a strange idea, but it's very similar to a toy top. Take a toy top and spin it. If you didn't have friction, the toy top would spin forever. This is very similar to how the inner magnetic ring of a magnet works.

The inner magnetic ring state of a tadtron is very similar to that of a tadtron in the electricity state. The only difference in the 2 states is that electricity is going to follow the medium that it is in. The inner magnetic ring is simply going to circle and stay inside the magnet. Magnetism easily travels though most matter. It must be very thin. The head on the tadtron must be very small. The tail on the tadtron in the inner magnetic ring state is probably quite long. This long tail will give us places to anchor the magnetic ribs. The difference between the electricity tadtrons and the magnetic tadtrons is a small difference in the head and the tail.

What is an inner magnetic ring? It is a large group of tadtrons following the circular path inside of a magnet. They circle until their state is changed to something else. If you have more circling tadtrons, then you have a stronger magnet. How many tadtrons are there in each inner magnetic ring? There could be hundreds, thousands, millions, billions or more. What is the diameter of the inner magnetic ring? The diameter could be microscopic or many miles across. It would depend on the magnet.

There are tiny magnets and big magnets like the earth. There may be multiple rings inside of rings. The black circle in the illustration is the inner magnetic ring. It is solid because of the large number of small circling tadtrons. All the tadtrons should be circling the same direction. This is the circular path of the tadtrons.

What is the purpose of the inner magnetic ring? It serves as a key ring, just like the molecule. The main difference is multiple tadtrons and a larger circle. The ribs now have a place to anchor and circle from. There are 15 purple circles with multiple diameters in the illustration. These are the magnetic ribs. They all go the same way. They circle out the top. This is the north pole of the magnet. When the tadtrons circle in the bottom it is the south pole of the magnet.

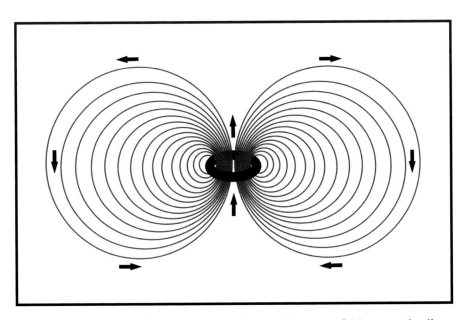

Illustration 18-6: 2D inner magnetic ring and 2 rows of 15 magnetic ribs

Cube magnet rings and ribs

Let's move to a 3D image to better understand the magnetic ribs. There are 2 illustrations. The first illustration is a picture of the top of a 1 cubic inch magnet. This is the side that is under the ferrofluid. It has a circle and an "N" on top of it. This is the north pole of the magnet. The circle is a simulated inner magnetic ring. We cannot see the inner magnetic ring or the magnetic ribs. Why? Light isn't reflected from them.

The second illustration is a 3D image of the top of a magnetic field. There is a single inner magnetic ring. There are only 37 rib rows illustrated, with 8 circles inside of each rib row. When the ferrofluid in the plastic dish is set on top of the cube magnet, what happens? Go back and look at the first ferrofluid picture. The ferrofluid follows the strongest magnetic ribs. Where are the strongest ribs? Obviously, next to the magnet. What then are the nubs or columns that form? Each column is following the circular path of a magnetic rib. The columns are only a fraction of an inch long. Why are they so short? The magnetic field becomes weaker as it continues on out for the full length of each magnetic ribs circular path.

Look at the illustration and imagine the ferrofluid on top of it. What would you see? You can see the mechanical workings behind a magnet. The magnetic ribs build the dome of ferrofluid. Imagine the second illustration on top of the first illustration. Imagine the ferrofluid going to the strongest magnetic fields. The mechanical workings of the magnetic field can be seen.

What are magnetic ribs? Magnetic ribs are tadtrons in the magnetic rib state. They have a job of circling through an inner magnetic ring. They stay in that state until they find something else to do. The magnetic ribs are very similar to electricity. They travel through most matter. They must be very thin. So, magnetic rib tadtrons must have a small head and a long tail.

Illustration 18-7: North pole top of cube magnet

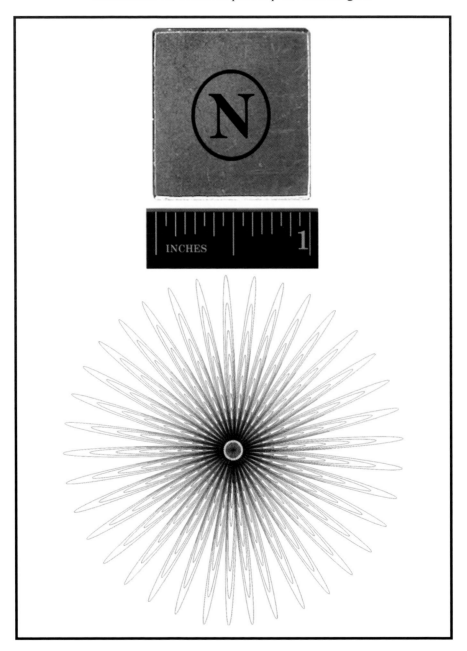

Illustration 18-8: 3D inner magnetic ring, multiple rows of magnetic ribs

Cross section of a magnet

The next illustration is a cross section of the inner workings of a magnet. Here is what you would see if we sliced the dome of ferrofluid in two. To do this, imagine taking a knife and cutting the ferrofluid. Then take what you have cut and lay it flat. We have taken a magnet and cut it from top to bottom. The top third of the illustration would be the cross section of the ferrofluid.

The bottom third of the illustration is the rest of the magnet. The blue magnetic ribs are the ribs that go at 90 degrees to the left and right of the inner magnetic rings. The black ribs at the top are ribs that circle out behind the magnetic spine. The black ribs at the bottom are ribs that circle in behind the inner magnetic rings. There is an unknown number of rows of inner magnetic rings. This is the magnetic spine that we will talk about later. The black line around the outside is the simulation of where the ferrofluid follows the ribs.

If you have less ferrofluid, what would the outline of the ferrofluid look like? Just follow a smaller row of ribs with your eyes. A dome with less ferrofluid would be almost identical to the first. The only difference is it would be smaller.

What would happen if we turned the magnet side ways under the ferrofluid? Before you see the answer look at this illustration. The north pole is at the top. The south pole is at the bottom. There are nubs or columns at the top. There are nubs or columns at bottom. The middle is a straight line. Imagine this in your mind. Now turn the page and see the answer.

Illustration 18-9: Cross section of ferrofluid

Magnetic ribs in ferrofluid

The next illustration is another picture of ferrofluid. Underneath the ferrofluid are the 2 magnetic cubes on their side. The north pole is to the top. The south pole is towards the bottom. You can see the magnetic ribs on the top where the north pole is. You can see the magnetic ribs at the bottom where the south pole is. Notice the cylinder between the ribs. The cylinder does not have any nubs or columns coming out of it. It is smooth. Why? We will explain that in the next illustration.

What are we seeing then? Remember how ferrofluid works. The ferrofluid follows the magnetic fields that are the strongest. The magnetic ribs at the north and south pole are obviously a strong field. The columns are more noticeable from this angle. It is easier to see the circular path that the magnetic ribs follow.

Why then is the cylinder in the middle smooth? This is where the magnetic spine will come into play. The next page will give this answer.

Illustration 18-10: Ferrofluid side with 2 cube magnet

Inner mechanical working of 2 cube magnets

This next illustration is the inner mechanical workings of the 2 cube magnets that are underneath the ferrofluid. We only used 1 size of circular paths of magnetic ribs. These ribs are about the same size as the cylinder that you see in the ferrofluid in the previous illustration. If we added in all the smaller ribs, the inside of the magnet would be almost impossible to see.

Look at the inside of the magnet. There are 9 inner magnetic rings. They are all lined up in a column named the Magnetic Spine. The magnetic spine acts just like the backbone in any mammal. It is the anchor point for all the ribs. The main difference between a magnet and a mammal is, the ribs on a magnet go out in all directions. The ribs of a mammal are held in place by muscles and tissue.

For simplicity we only show 9 inner magnetic rings in the magnetic spine. If you add more than 9, you cannot see the spine. Look around the outside edge of the ribs. What do you see? You can see the same thing as what is in the ferrofluid. The left and right sides of the ribs are smooth. The top and bottom of the ribs have nubs or columns. The ferrofluid forms on one set of ribs. If you add more ferrofluid it may form the next larger size of ribs. If you take away some of the ferrofluid it will form around a smaller set of ribs.

Look at the right and left sides of the ribs. Remember these ribs are the circular path of tadtrons. They are all spinning out of the top center and circling back in. Along the sides you have tadtrons coming out and going in. This will account for why the ferrofluid is so smooth. The ferrofluid cannot follow any particular line or rib of tadtrons because they are going in 2 direction. However, at the north and south poles, the ferrofluid is able to follow a rib of tadtrons because there are no other tadtrons going in a different directions.

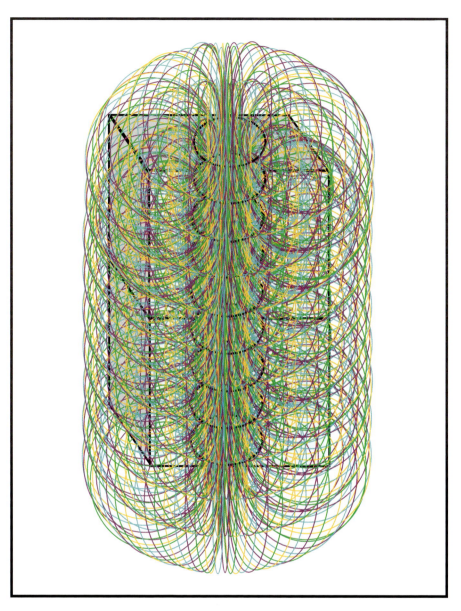

Illustration 18-11: Side view of the 2 cube magnets

Magnetic field and gravity effect ferrofluid

The next illustration is another picture of ferrofluid. The ferrofluid has the washer magnet underneath it. The round north pole side of the magnet is facing up. What do we see? We see the same dome of ferrofluid as we saw with the magnet that had the 2 cubes. What conclusion can you get from this? The inner magnetic rings, the magnetic spine and the magnetic ribs are the same regardless of the medium!

This illustration shows the top side of the washer magnet. This magnet is not as strong as the 2 cube magnets. You can see some excess ferrofluid around the sides of the dome. Ferrofluid follows the strongest magnetic fields. Gravity plays a role in the shape of the ferrofluid. The ferrofluid gathers at the bottom or base of the dome. It is being affected by both gravity and the magnetic field.

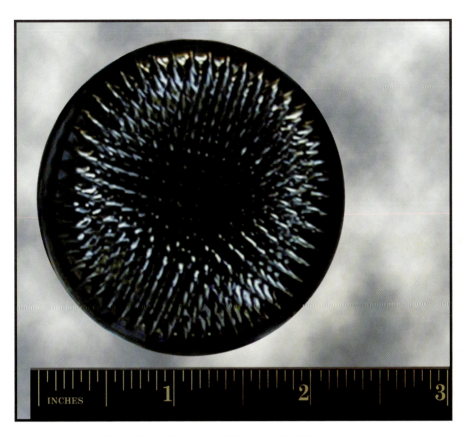

Illustration 18-12: Picture of ferrofluid with washer magnet

Washer magnet rings and ribs

The next 2 illustrations show what's under the ferrofluid. The first illustration is a picture of the top of the washer magnet. This is the side that is under the ferrofluid. It has a circle and an "N" on top of it. This is the north pole of the washer magnet. The circle is a simulated inner magnetic ring.

The second illustration is the same as the one with the cube magnet. The main difference is the outline of the washer magnet. The red lines that appear to be bent are an optical illussion. The inner magnetic ring, the magnetic spine and the magnetic ribs are the same. They are simply in a different magnet.

Illustration 18-13: Picture of top of washer magnet

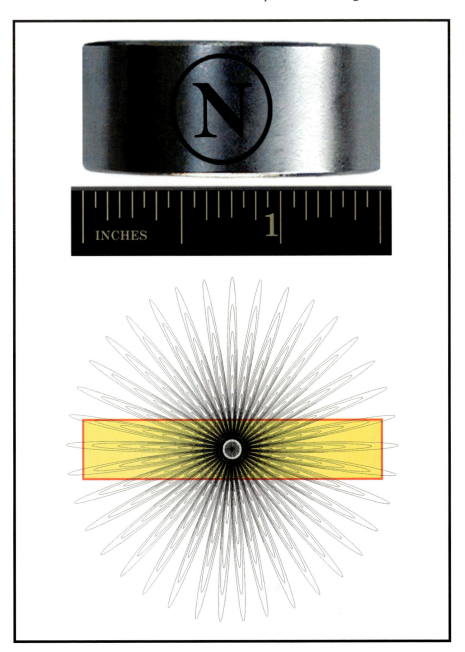

Illustration 18-14: On top of washer magnet

Ferrofluid with washer magnet

This illustration is another picture of ferrofluid. The washer magnet is under the ferrofluid. The washer magnet is laying flat on its side. The washer magnet is only one and a half inches across. Notice how similar this looks to the ferrofluid with the 2 cube magnets with ferrofluid. There is difference in the length of the magnets. This will account for why they look differently. The washer magnet is shorter.

If you compare the washer magnet to the 2 cube magnets you will notice these things. The ends look almost identical. You can see the circular path of the magnetic ribs. The center is the smooth cylinder shape. The only difference is the length of the smooth cylinder.

There is one other thing that seems interesting. There is a hole in the washer magnet. The hole is under the smooth cylinder. The hole does not effect the magnetic ribs.

Illustration 18-15: Picture of ferrofluid with washer magnet under it

Inner mechanical working of washer magnet

The next illustration is the inner workings of the washer magnet. There are only 7 inner magnetic rings in the magnetic spine. We used 2 less inner magnetic rings to account for the half inch difference between this magnet and the 2 cube magnets. The magnetic spine is able to continue through the hole in the washer magnet.

If you follow the outside edges of the magnetic ribs you will be able to see what is happening in the ferrofluid. It is smooth on the sides. It has nubs or columns at the north and south poles.

I chose 2 different types of magnets to do a comparison. I did this so you can see that the magnetic spine and the magnetic ribs behave the same, even though the magnets have different sizes and shapes. If you use a horseshoe magnet, what would be different? The spine would just bend to follow the magnet. Everything else will be the same.

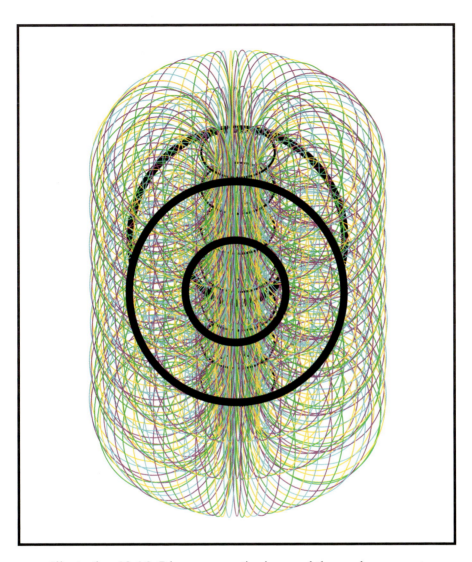

Illustration 18-16: 7 inner magnetic rings and the washer magnet

Magnet effect on iron fillings

The next illustration is a picture of iron filings on a piece of paper. Underneath the paper is some cardboard and a small Neodymium Iron Boron magnet. The Neodymium Iron Boron magnet is a rectangular magnet. The magnet's dimensions are 1 inch by 1/4 inch by 1/4 inch. The paper is just a standard 8 ½ by 11, white sheet of paper. The iron filings I made by grinding an iron nail.

The north pole of the magnet is facing to the top of the page. I sprinkled the iron filings all over the paper. The iron filings go to the points that the magnetic field is the strongest. Some cool looking patterns start to develop. There are many irregular lines that form. There are some oval shaped lines that form on the sides of the magnet. The north and south poles have many lines that circle out.

In the past these patterns of lines were called the magnetic lines of force. What's behind these patterns? The next illustration will give the answer.

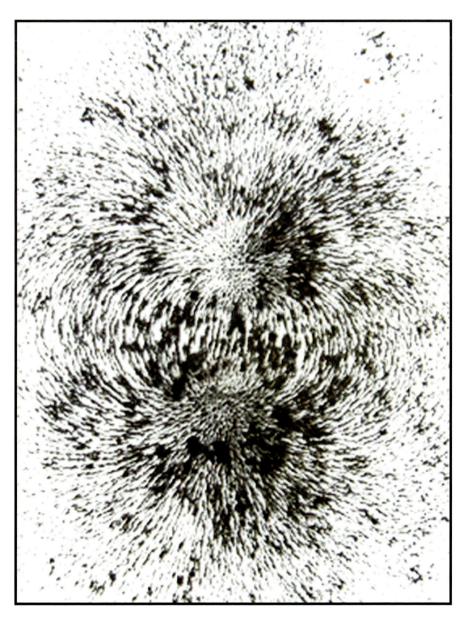

Illustration 18-17: A picture of iron filings

Magnetic lines of force

This next illustration shows the inner workings of a magnet. There are 9 inner magnetic rings in the magnetic spine. There are 15 ribs coming out of each inner magnetic ring. The magnetic north is at the top of the page. All the magnetic ribs are blue.

What I am going to do now, is to show how the magnetic lines of force are formed. Imagine putting a piece of paper over this magnet. Then sprinkle some iron filing on the paper. Where will the iron filings go? They will go to where the magnetic force is the strongest. Where will that be? It will be where the most magnetic ribs intersect. We did this with the black lines. We tried to follow where ribs intersect and on the outer edges of ribs. The magnetic lines of force are somewhat oval shaped. We only followed those paths for simplicity. There are lots of other intersections. The iron filings will go to those intersections. Go back and look at the previous illustration. You can see these other intersections. Also remember there are other things that are happening too. The picture takes into account, many 3D effects.

Look at this illustration again. The black dashes follow the first 8 circumferences of magnetic ribs. Also, additional ribs cross the rows, which produces a concentrated magnetic field. More iron filings are attracted to these crossings. Look at each larger rib crossing. What do you see? You can see the same shape that forms in the ferrofluid and the iron filings. It takes the shape of a pill or a capsule of medicine.

This explains the magnetic fields. Next we are going to show how magnets pull and repel.

Illustration 18-18: Lines of force

Basic principals of how magnets work

The next illustration is to show how magnets produce the pull effect. Anyone, that has worked with magnets, knows they pull together.

To explain how this works, let's use an example of me and Uncle Mike standing 10 feet apart. If I throw a baseball at Uncle Mike and I hit him, he will move backwards from the force of the baseball. If Uncle Mike throws a baseball and it hits me, I will move backwards. When the ball or particle hits the other person, they will be moved. If me and Uncle Mike each throw a baseball at the same time, and the baseballs collide in the air, nothing will happen to either me or Uncle Mike.

If I throw a boomerang and it circles around and hits Uncle Mike in the back, he will be moved towards me. If Uncle Mike throws a boomerang and it circles around and hits me in the back, I will be moved towards him. The more boomerangs we throw the more we will be "pulled" towards each other. If the boomerangs hit in the air nothing happens to me or Uncle Mike. If the boomerangs were transparent, what would this look like from a distance? It would look like me and Uncle Mike were being pulled together.

Lets add another variable. The wind is blowing. The boomerangs don't fly in a perfect circle but they still hit me or Uncle Mike in the back. What is the difference? Nothing. It's where the boomerangs hit that causes me or Uncle Mike to move. If the Boomerang hits me in the front, which way will I move? I will move backwards.

There are 2 illustrations. The first is a stick man throwing a boomerang that catches the wind and hits the inside of a hoop. The hoop is pushed forward towards the stick man.

The second illustration is another stick man throwing a boomerang. This time the boomerang catches the wind and hits the outside edge of the hoop. The hoop is pushed away from the stick man.

Both these illustrations show the basic principles of how magnets work.

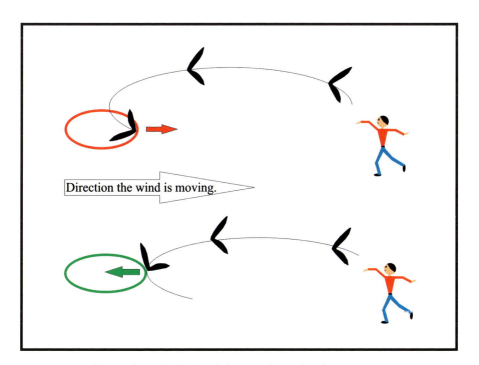

Illustration 18-19: 2 stick men throwing boomerangs

The pull (or attract) effect

Magnetic ribs work the same as the boomerangs. Imagine each circling tadtron in a magnetic rib as a boomerang. They circle around and hit the inner magnetic ring of another magnet. That inner magnetic ring is part of the entire magnetic spine. When the tadtron hits, it will cause the entire magnet to move in the angle of the hit.

Look at the next illustration. There are 4 pictures of 2 magnets at various distances apart. This shows a cross section of the magnet field of the 2 magnets. Each magnet has 1 inner magnetic ring and 1 set of magnetic ribs. There are tadtrons circling on both sides. The tadtrons are circling in the same direction. The blue tadtrons are magnetic ribs for the left magnet. The red tadtrons are the magnetic ribs for the right magnet. They are different colors so you can tell which go with which.

The left magnet has the north pole to the top. The magnet on the right has the south pole to the top. The circling tadtrons in the middle will cause interaction between the magnets. The north and south are turned this way so that the tadrons in the magnetic ribs are circling in the same direction.

When a tadtron from 1 magnet circles inside another magnets inner magnetic ring, what happens? It puts pressure on the inside of the opposite inner magnetic ring. The whole magnet is then pushed in that direction. If you have more tadtons connecting, you will have more pressure. This produces the Magnetic Boomerang Pull Effect. As the inner magnetic rings get closer, more tadtrons produce a stronger Magnetic Boomerang Pull Effect. The magnets are the black squares. As they get closer, follow the paths of the tadtrons. The tadtrons paths are deflected as they enter the path of an opposing magnetic rib. The tadtrons are following the path of least resistance. It's like following water into a drain.

Look at each set of magnets. It's easy to see that as the magnets get closer more tadrons produce more pressure or pull. What's the result? Magnets that are put side by side with the north and south facing different directions pull each other together. That's how they work. Get some magnets and try it.

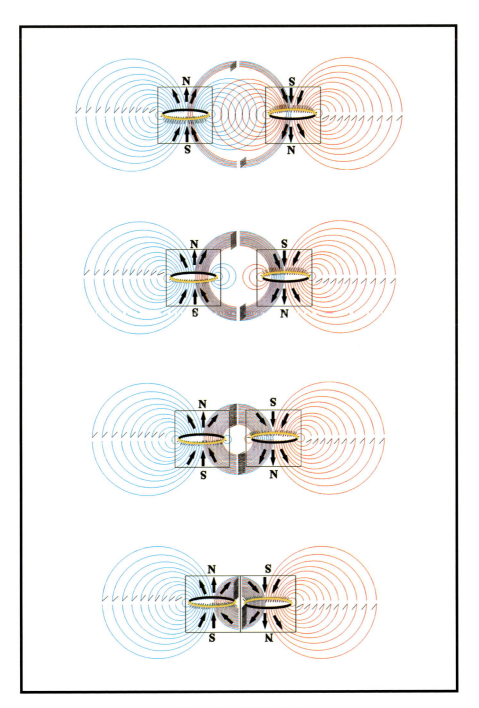

Illustration 18-20: Pull

The push (or repel) effect

The next illustration is the opposite of the last illustration. Now we are going to show why magnets push apart or repel each other. Again we have 4 sets of magnets. This time the north pole is to the top on both magnets. These illustrations are 4 cross sections of the inside of 2 magnets.

Now the tadtrons in the magnetic ribs are circling in opposite directions. This will make a big difference in what will happen. Look at how the tadtrons come out of the top side. It's like water coming out of the end of a hose. When water is coming out, you cannot force water back in that end of the hose.

What will happen when the tadtrons interact? They cannot go in an opposing inner magnetic ring. They must pass to the outside of the opposing inner magnetic ring. When they pass to the outside, what will they do? They will apply pressure on the outside of the inner magnetic ring. The pressure will cause the magnet to be pushed away or repelled. I will name this the Magnetic Boomerang Repel Effect.

Look at each of the 4 illustrations. As the magnets are forced closer, you can see the pressure build. As the magnets are forced closer, more tadtrons interact. The pressure will be proportional to the increased number of tadtrons.

Take 2 magnets and play with them. Hold them like the illustration. Push them together. Feel how they repel. As you push them closer, it gets very hard to hold them together.

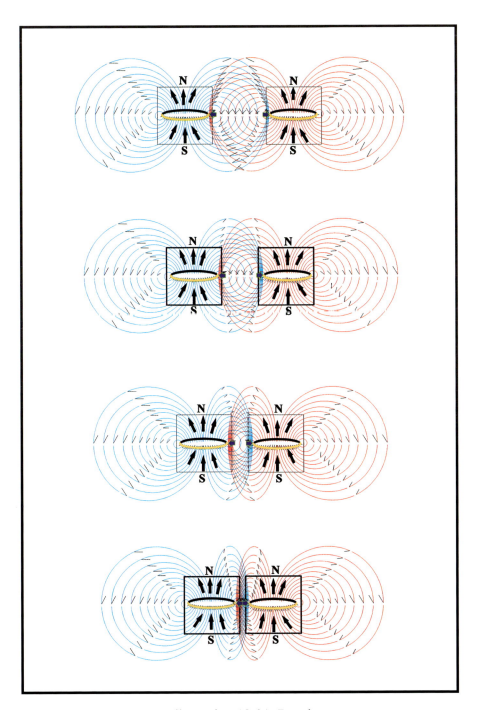

Illustration 18-21: Repel

Magnetic boomerang pull and repel effect

The next illustration has 2 sets of magnets. The top left magnet has the north pole facing up. It also has tadtrons in blue. The opposing magnet is on the top right. That magnet has the south pole on the top with tadtrons in red. There are 7 inner magnetic rings in the magnetic spine of each magnet.

How does the top pair of magnets work? They pull each other together. The opposing tadrons circle in and produce the pressure, inside the spinal ring, that causes the Magnetic Boomerang Pull effect. Now there are more inner magnetic rings. What does that do? It makes a stronger magnet. Also, the magnetic spines try to align themselves. Both magnets try to turn parallel to each other. Why? They are going to what is the strongest pull. The top and the bottom are pulled to the top and bottom of the opposing magnet. We will talk about this alignment on the following pages.

Next lets deal with the bottom pair of magnets. Both magnets have the north pole at the top. Each magnet has 7 inner magnetic rings in their magnetic spine. The bottom left magnets have tadtrons in blue. The bottom right have tadtrons in red. What do these magnets do? They repel each other. Why? It's the Magnetic Boomerang Repel Effect. Look at all the tadtrons that are producing pressure. What is the result of having 7 inner magnetic rings? You have a very strong Magnetic Boomerang Repel Effect. Look at the tadtrons that are deflecting off of the opposing inner magnetic ring.

Get 2 magnets that are long and shaped like these. Face the poles like the top set of magnets. Bring the magnets together. You can feel the Magnetic Boomerang Pull Effect. Turn 1 magnet over so it is like the bottom set of magnets. Bring the magnets together. Now you can feel the Magnetic Boomerang Repel Effect.

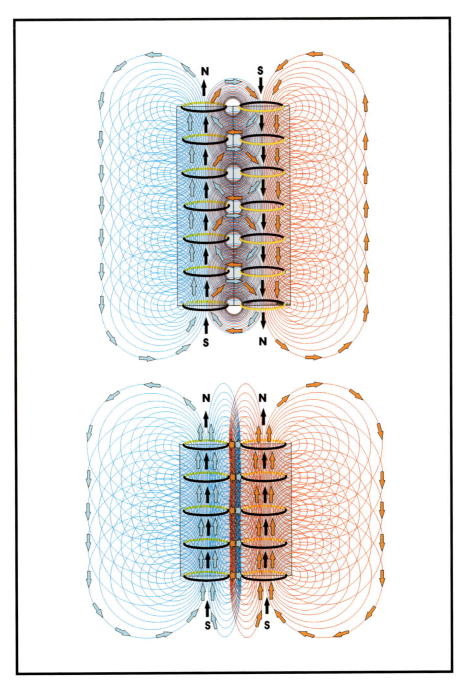

Illustration 18-22:
2 sets of magnets with 7 inner magnetic rings in spine

How a compass works

The next illustration is to show how a compass works. Magnets will work the same no matter what their size is. In this illustration we have drawn an example of the earth and a small compass.

Look at the picture of the earth. There is a magnetic spine through the center of the earth. We have some magnetic ribs coming out of the magnetic spine. How big are the inner magnetic rings? I don't know. They are probably many miles across. How far out do the magnetic ribs extend? I don't have an exact figure. We know they extend for several thousand miles or else compasses wouldn't work.

Magnetic north is not aligned with the north pole but with the earth's magnetic spine. It is really the south pole of the earth's magnetic spine. This is where all the compasses point to. The earth's magnetic field is considered to be very weak.

Look at the illustration. The earth is a big magnet. The compass is a little magnet. Set the little magnet on the big magnet and what happens? The magnetic spines try to align themselves. The little magnet will do the alignment because it can't move the big magnet. Look at the illustration. Notice a magnetic rib, from the earth, pulling at the top inner magnetic ring in the compass. Notice another magnetic rib, from the earth, pulling at the bottom inner magnetic ring in the compass. This pulling effect is what causes the compass to align itself parallel with the earth's magnetic spine.

How does this work? Think of 2 people standing on a dock. They each have a rope that hooks to a boat. One rope is hooked to the front of the boat. One rope is hooked to the back of the boat. Both people pull equally. What happens? The boat will be pulled sideway towards the 2 people. The boat will stop when it gets to the dock. The boat will be parallel to the dock and the 2 people. This is the principle that magnets work on. The big difference is there are more ropes pulling (Magnetic Ribs) and more places that the ropes connect to (Inner Magnetic Rings).

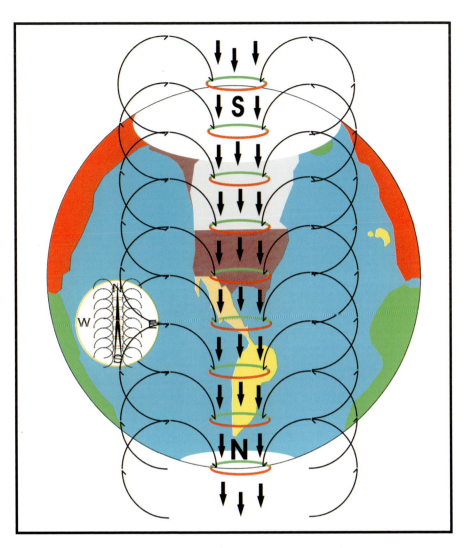

Illustration 18-23: How a compass works

Pull and repel effect when poles are perpendicular to one another

In this illustration we are going to show the pull and repel effect when the poles of the magnets are perpendicular to one other. The next illustration has 2 sets of magnets. Let's work with the top set first.

The north poles of both magnets are facing to the top. This means all tadtrons are circling in the same direction. What does that do? It allows the tadtrons to be able to circle through the opposing magnets inner magnetic ring. What is the result of that? It produces the Magnetic Boomerang Pull Effect. In the previous illustrations we showed the pull effect from the sides of the inner magnetic rings. Now the pull is from the top and bottom. What's different about that? The spines are aligned end to end. The pull is top to bottom not side to side!

Look at the top illustration. The purple circles are the shared circling tadtrons. See how they produce a pull. Where do the magnets want to align? At the strongest point. Where is that? Inner magnetic ring to inner magnetic ring.

Look at the bottom set of magnets. The poles are now opposite to each other. What does that mean? Tadtrons are circling in opposite directions. What does that mean? Tadtrons can't go through an opposing magnets inner magnetic ring. What is the result? It produces the Magnetic Boomerang Repel effect. The tadtrons now strike the top or bottom edge of the opposing inner magnetic ring. What does that do? It causes the magnets to be repelled up or down. Quite simply, they push the other magnet away. Now you see how the pressure builds between the 2 magnets. It's easy to see why magnets repel on opposite poles. The little arrows show where the pressure builds.

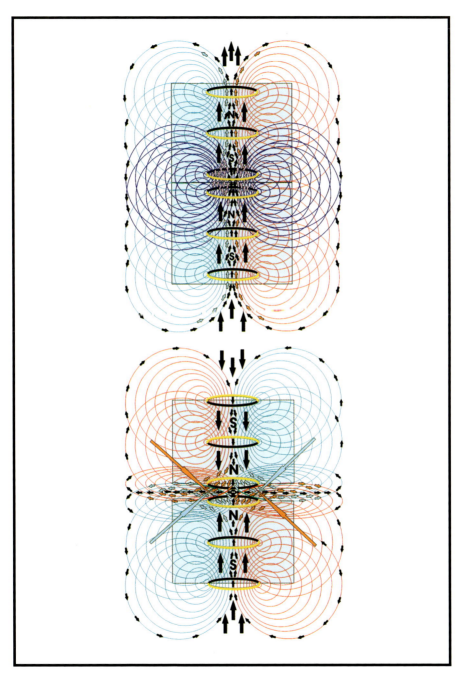

Illustration 18-24:
Pull and Repel effect when poles are next to each other

How magnets attract iron

How do magnets attract iron? There's a very simple answer. The tadtrons from a magnet charge the iron and turn it into a magnet. It's just like an electro magnet. When you make an electro magnet, you cause a magnetic spine to be created inside of the iron. The magnetic ribs then form and now you have a magnet. What does a magnet do to a piece of iron? It causes a magnetic spine to form in the iron. The magnetic spine then produces the magnetic ribs. The spine may form vertically or parallel to the magnet. We have three examples in the next illustration of a magnet and 2 iron nails. The blue magnetic ribs circle counter clockwise. The red magnetic ribs circle clockwise.

The first example has the magnet above two nails. A magnetic spine forms through the nail. The spine has formed vertically. The magnetic ribs form. The nail now has all the features of a magnet. The nail is able to attract a second iron nail.

The second example is a magnet attached parallel to the nail. A parallel magnetic spine forms in the iron nail. The magnetic ribs form. The nail now has all the features of a magnet. The nail is able to attract a second iron nail. The vertical spine seems to be stronger than the parallel spine.

In the third example we pried the magnet away from the nail. When the magnet reached the angle shown, the magnetic field in the nail broke and the bottom nail fell off. You can do this experiment at home.

What is going on? The tadtrons from the magnet leave and become a magnetic field in the iron nail. The magnet doesn't lose any of its strength as it charges the iron. So, if tadtrons leave a magnet and go to the iron, where do the replacement tadtrons come from? We will answer that in Law 19.

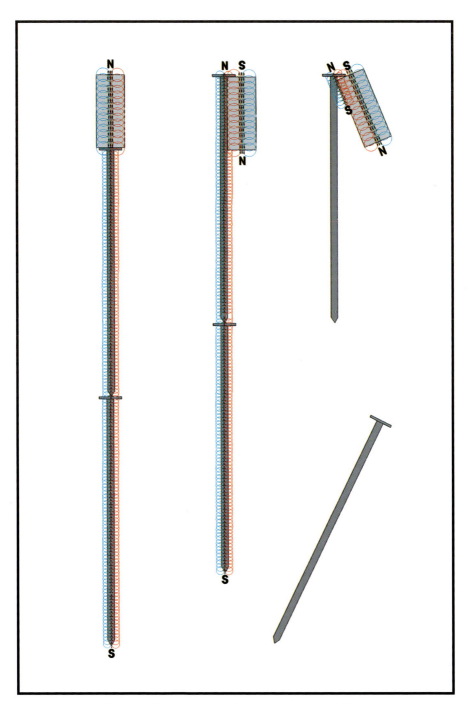

Illustration 18-25: Ways that magnets attract iron

Law 18: Closing comments

There is one other phenomenon about magnets I would like to talk about. It is diamagnetism. There are materials that are diamagnetic. Diamagnetic is the opposite of magnetism. Diamagnetism repels instead of attracting. Bismuth is a diamagnetic material. Superconductors are diamagnetic. How does diamagnetism work? Just the opposite of iron. If the iron nail in the previous illustration was bismuth or it was a superconductor, the magnetic ribs would rotate in the opposite directions. The materials will repel each other rather than attract.

When a magnet is moved by a metal object, a current of electricity is produced in the metal. That's how generators work. Generators never lose their magnetism and they run for years. We will explain that in Law 19.

Once you understand the magnetic spine and the magnetic ribs they are quite simple. This is one more piece of the puzzle that is turned in the right direction.

Law 19

Law 19

Exiting gravity changes state after passing through the atom

Law 19: Opening comments

This is part 4 of the gravity section of the jigsaw puzzle. This law will wrap up most of my theory. Law 19 may be the most important of all the laws. You will need to read all the other laws to understand this part. I will be answering a question that should have been asked many years ago. What happens to gravity after it gets to it's atom?

These are the questions I am going to answer in this law. Where does the energy in sound come from? Where does the energy in waves in water come from? Where does the energy from magnets come from? Where does infrared light come from? When something is heated and it gives off light, where does that light come from? What heats the center of the earth? What powers the sun?

The answer is going to be very simple. This is the balance of the universe. If something goes in then something must come out. It must be equal. Whatever amount of tadtrons go in, is the same amount of tadtrons that come out. What goes in may be one state. What comes out may be another state. The amount of tadtrons must be the same. Personal gravity is what is going in. Let's move on and talk about what is coming out.

Exiting gravity can be changed to sound

Where does sound come from? The illustration on the next page will give an answer. The illustration is a cross section of a drum. There are 2 drumsticks above the drum. One drumstick has moved down and struck the head of the drum. Sound is the result. We are going to explain what's behind the sound.

Slightly above the drumstick is an atom. It represents the atoms in the drumstick. Slightly below the head on the drum is another atom. It represents the atoms in the drum head. The blue arrows above the drum represent the personal gravity of the drumstick. When the drumstick hits the drumhead, the exiting gravity tadtrons are mashed or changed into sound tadtrons. The sound tadtrons are now corkscrews. They are coming out the bottom of the drum.

This is how sound is made. When atoms collide, the exiting tadtrons change to sound. The sound that comes out will depend on the size and shapes of the objects. How hard the objects collide will affect the sound. If a bowstring is pulled over the strings on a violin, sound is the result. Any collision or any friction produces sound.

Once the corkscrew tadtrons are produced, they will continue on as sound until they change states to something else. You can beat on a drum as long as you want. You have an unlimited supply of sound tadtrons.

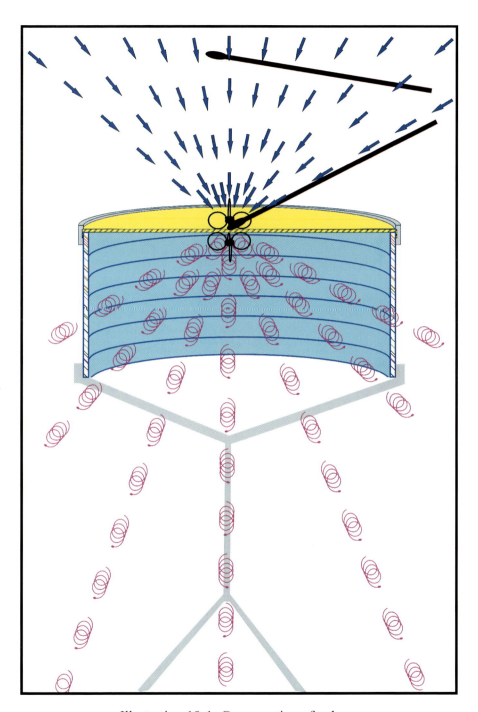

Illustration 19-1: Cross section of a drum

Exiting gravity can be changed to slow sound

The illustration on the next page is an aluminum boat traveling through the water. When a boat travels through the water it produces waves. The waves have energy in them. Slow sound was explained in a previous law, by putting tadtrons in the waves. This is the source of those tadtrons.

Above the boat are arrows representing the personal gravity of the boat. An aluminum molecule at the front of the boat represents all the aluminum in the boat. What happens when the aluminum comes in contact with the water? The exiting gravity from both the aluminum and the water are changed to slow sound. The collision or contact causes corkscrews to form. The size, shape and direction of the corkscrews depend on the speed and shape of the boat. Different boats make different waves.

What else makes waves? Anything that comes in contact with water will make a wave. The principle will be the same. When an object comes in contact with water, the exiting gravity will change to slow sound corkscrew tadtrons. The size of the wave will depend on the size, shape, speed, and weight of the object.

Wind makes waves in water. If you have a constant wind blowing in the same direction, bigger waves will form. If you have a wind that changes its speed you have choppy waves. What's the difference? A constant wind will produce corkscrews at the same speed, forming big waves. Winds that change speed will generate corkscrews that have different speeds and shapes, thus making the choppy waves.

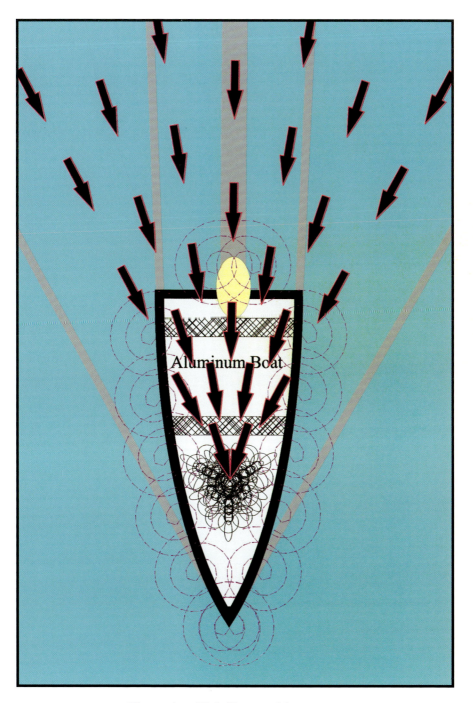

Illustration 19-2: Boat making waves

Exiting gravity powers magnets

On the next page is an illustration of a horseshoe magnet. At the top of the magnet is an atom. This is a representation of all the atoms in the magnet. At the bottom of the magnet are 2 inner magnetic rings. Each inner magnetic ring has a full set of magnetic ribs. This is a representation of the entire magnetic spine and magnetic ribs. If you show a full spine, you will not be able to tell what's going on.

Above the magnet are black arrows representing the personal gravity of the magnet. The atom has gravity passing through it. When the gravity exits the atom, it is now looking for something to do. The arrows coming out become part of a magnetic rib or part of a magnetic spine. They can fill in any gap in the magnetic spine. There are an unlimited number of replacements.

When you take a magnet and touch it to an iron nail, the iron nail becomes magnetic. Where do the tadtrons for the magnetic spine and ribs come from? They come from the exiting gravity. A magnet can make an iron nail into a permanent magnet. A single magnet could charge millions of nails and never lose any of its own magnetic strength.

When you run a generator, electricity comes from a magnet being passed by metal wires. Magnetic rib tadtrons change to electrons. Exiting gravity tadtrons replace the magnetic rib tadtrons. A generator can run forever. The supply of tadtrons will never be depleted.

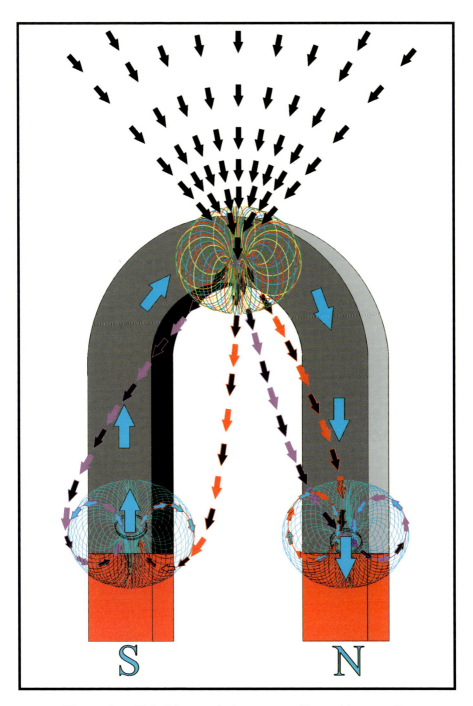

Illustration 19-3: Magnets being powered by exiting gravity

Exiting gravity produces infrared light

We are showing where infrared light comes from in the illustration on the next page. The source of the infrared light will be the man. We have an atom sitting over the middle of the man. The atom represents all the atoms and molecules inside the man.

Above the man is a bunch black arrows representing the personal gravity of the man. The personal gravity will travel through all the atoms in the man. When the gravity exits an atom, it will be looking for something to do. The instruction will be based on how hot the atom is. A certain percentage of exiting gravity tadtrons will change to infrared light tadtrons. If the atom is hotter, more exiting tadtrons will change to infrared light tadtrons. If the atom is colder, fewer exiting tadtrons will change to infrared light tadtrons. The infrared light tadtrons are coming out the bottom of the man.

It takes heat sensing equipment to view infrared light. If the man was walking in a cold place, he could be viewed, even in total darkness, with this equipment. Everything gives off infrared light. The special equipment can gauge how hot some things are by how much infrared light is given off. If you were cold and you put your hand up next to the man, you could feel heat coming from him. It is the infrared light that you feel.

We have gravity going in and have infrared light tadtrons going out.

Illustration 19-4: Infrared light coming from a man

Exiting gravity produces red light from a red-hot bolt

The illustration on the next page is a red-hot bolt. We will be showing where red light comes from. The source of the red light will be the bolt. We have an atom sitting over the middle of the bolt. The atom represents all the atoms and molecules inside the bolt.

The black arrows represent the personal gravity of the bolt. The personal gravity will travel through all the atoms in the bolt. When the gravity exits an atom, it will be looking for something to do. The instruction will be based on how hot the atom is. A certain percentage of exiting gravity tadtrons will change to infrared light tadtrons. We saw that in the last illustration. At extreme high temperatures the exiting gravity tadtrons will change to red light tadtrons. The red light tadtrons are coming out the bottom of the red-hot bolt.

You don't need special equipment to see this. If you heat the bolt hotter, it will turn a cherry red. If you heat the bolt even more, it will turn yellow and almost white. What is happening? Some of the exiting tadtrons are changing to green and blue light tadtrons. As long as the bolt stays hot enough, it will keep producing visible light tadtrons. It will also be producing infrared tadtrons. You will be able to feel these with your bare hands. Within a few inches of the bolt, you may get burned.

We have gravity going in. We have visible light and infrared light going out.

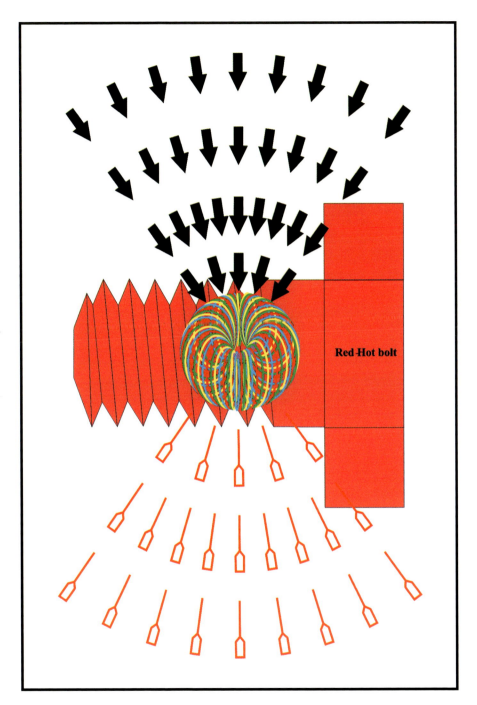

Illustration 19-5: Red-hot bolt

Exiting gravity heats the center of the earth

What heats the center of the earth? This illustration on the next page is an attempt to explain this. This is a cross section of the earth. At the center is an atom that represents all the atoms and molecules of the earth.

The black arrows at the top represent the personal gravity of the earth. The center of the earth is a concentration of gravity. What is at the center of the earth? No one really knows. It is believed to be very hot. It is believed to be many times hotter than the red-hot bolt. What would the exiting gravity tadtrons be changing to? They would change into infrared light tadtrons, visible light tadtrons and some unknown tadtron states. What's the result? The center of the earth stays very hot.

You can see the tadtrons coming out of the bottom. They cause additional heat, change state, and are looking for something to do. The excess heat comes out in volcanic eruptions. We have shown 3 volcanos in the illustration.

Several types of radiation are known to come out of the earth. Gravity goes in. It heats the center of the earth. Lava comes out through volcanoes. Radiation comes out of the earth. Tadtrons in an energy state yet unidentified come out. Something powers the earth's magnetic field. There are plenty of tadtrons for that. What about the aurora borealis? They are known as the northern lights. What powers them? Exiting gravity from the earth. We know gravity goes in, we know many things come out.

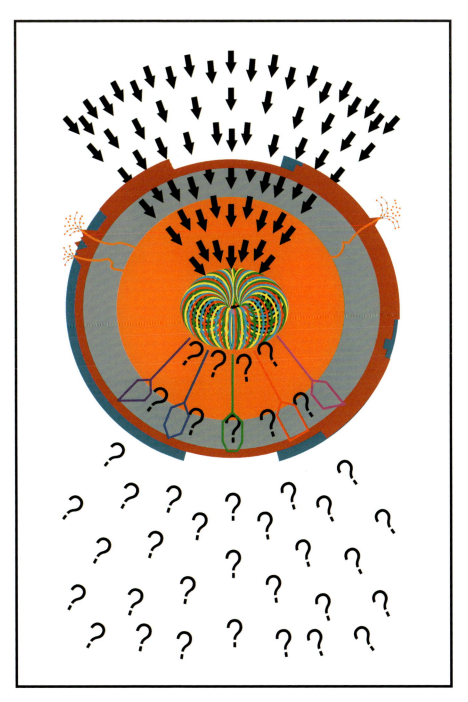

Illustration 19-6: Exiting gravity heating the center of the earth

Exiting gravity powers the sun

What powers the sun? This has been a mystery for many years. There is an easy answer based on everything we have covered. The illustration on the next page is a cross section of the sun. There is an atom at the center. This atom represents all the atoms and molecules in the sun.

The black arrows at the top represent the sun's massive amount of personal gravity. The sun is very hot. What happens to exiting gravity when it goes through a hot atom? It is converted to infrared light, visible light and in this case ultraviolet light. It's all the things coming from the sun. Every day the sun heats up the earth. Massive amounts of energy come from the sun. We are showing radio waves coming from the sun, and unknown energy as question mark.

Gravity goes in. Visible light comes out. Infrared light comes out. Ultraviolet light comes out. Radio waves come out. Unknown energies come out. Do you know what this means? THE SUN WILL RUN FOREVER! It will never burn out. It's built to keep on going and going and going.

The sun is a gravity converter. Can this be reproduced in the lab? When this day comes, we will have all the energy we need.

Illustration 19-7: Exiting gravity producing energies coming from the sun

Law 19: Closing comments

Exiting gravity has always been there. It's never properly been identified. Exiting gravity is a huge source of energy. It's clean. It's renewable. We need to learn to use it.

Law 20

Law 20

Putting it all together

Law 20: $U = \#Td$

This jigsaw puzzle is together. Now let's stand back and look at it. What do we see? Something that is not that hard. It's quite simple when things are arranged in the correct place. It's a universe full of tadtrons. Tadtrons make up everything we see, hear, smell, taste or touch. How many tadtrons are there. It's a number beyond my ability to comprehend. I will call the number of tadtrons in the universe a tadtrillion. Most people need a mathematical equation. The new equation is $U=\#Td$. U stands for the Universe. The # sign represents the number. Td stands for tadtrons. That's the universal equation. The universe equals the number of tadtrons. I could also put it this way, $U=\#Td=1$ Tadtrillion Tadtrons.

Let's go on a journey. Let's start with a tadtron and follow it through the entire spectrum of known and unknown energy states. Let's give our tadtron a number of 101. Tadtron 101 is out at the end of the universe. I am going to give Tadtron 101 a nickname. His new nickname is T101. He is looking for something to do. He finds the end of a gravity chain. He gets in line with a gravity chain and follows it. He travels for millions of light years. He goes through the galactic super mass of the milky way galaxy. He causes a small push on the galactic super mass as he passes through. This small push helps hold the milky way galaxy in orbit of the universal super mass.

T101 continues on until he reaches the universal super mass. Once he's there, he goes through his atom. He converts to infrared light. He now heads in a straight line for earth. T101 travels many light years. He travels through our atmosphere. He goes in an infrared telescope. The telescope converts the infrared light tadtron to a red light tadtron. The red light tadtron goes to a scientist's eye. The red light tadtron is converted to an electric signal which goes to the scientist's brain. The scientist's brain tells the scientist there is something red out there. The scientist doesn't know what he is looking at. It's different than anything else he has looked at.

T101 now converts to an exiting brain wave and leaves the scientist. T101 hits a wall and is converted to an energy source that is looking for something to do. Our tadtron travels to the end of the milky way galaxy. He finds the end of a gravity chain. He follows it. T101 travels many light years. He goes through our sun. He causes a small push. This small push helps hold the sun in orbit of the galactic super mass. T101 continues on until he comes to the galactic super mass at the center of the milky way galaxy. He passes through the atom he is going to. He converts to a blue light tadtron. He now heads back to our earth. T101 goes into a telescope. The blue light tadtron goes to a scientist's eye. The blue light is changed to an electron, which goes to the scientist's brain and tells him that he is seeing light. The scientist thinks he is looking at a super massive black hole. The only problem is light shouldn't be able to escape a black hole. The scientist is troubled. T101 gets converted to a brain wave and exits the scientist's mind. T101 is now looking for something to do.

T101 finds the end of a gravity chain at the edge of our solar system. He changes state to a gravity stream and follows the gravity towards our sun. T101 travels through our earth. It produces a slight push, which helps hold the earth in orbit around the sun. Our tadtron now continues on until it reaches the atom it is going to at the center of our sun. T101 gets converted to a green light tadtron and heads to earth. He strikes a green leaf on a corn plant. T101 is reflected to a farmer's eye, which converts it to an electron and sends it to the farmer's brain. The farmer sees a green corn leaf. T101 gets converted to a brain wave, leaves the farmer's brain, hits a corn stalk and gets changed to an unknown energy state.

Our tadtron moves past the moon and finds another gravity chain. It passes through the moon and produces a slight push that helps keep the moon in orbit around the earth. T101 goes to the center of the earth and goes through its atom. He gets converted to an infrared light tadtron. It strikes another tadtron at the center of the earth. That tadtron heats up. T101 changes to an unknown state and heads out the north end of the earth.

Our tadtron hits some solar flare material and becomes a red light tadtron that heads towards the eyes of someone in Alaska. After entering the eye, T101 gets changed to an electron and goes to the Alaskan's brain. He sees the Northern Lights. Our tadtron continues out from the brain, slams into an iceberg and changes to a tadtron looking for something to do.

He does a sun cycle, to the sun as gravity and comes back as a green light tadtron. T101 hits a solar panel and gets converted to an electron. T101 moves through the electricity grid to a radio station. Our tadtron gets converted to a radio signal and travels to a radio. The radio picks up the signal and he gets converted to another electricity tadtron. T101 travels to a magnet in a speaker and gets changed to a sound tadtron. Our sound tadtron travels to the ear of a person listening to the radio. T101 vibrates the person's eardrum and he hears the sound. T101 gets converted to an unknown state.

He is floating around and gets caught in a thunderstorm. Our tadtron converts to a lightning tadtron. Then it is converted to a red light tadtron and strikes a tree leaf. T101 is then converted to an unknown energy source.

T101 finds a gravity chain and follows it. Mr. T101 is now going through an atom in a magnet. He comes out of an atom and finds an empty space in an inner magnetic ring. Our tadtron then changes states to a magnetic rib. The magnet is part of a generator, that is in operation. The magnet is going by a copper coil. T101 hits the coil and gets converted to an electron. He is now back in the electric grid. He travels thousands of miles through copper wires. T101 goes through an electromagnet. He becomes a magnetic rib again. He circles and hits the inner magnetic ring in a piece of iron that is being lifted by the electromagnet. T101 causes the magnetic boomerang pull effect. It helps hold the piece of iron. T101 now changes to an unknown state looking for something to do.

He finds a gravity stream that is going to an oxygen atom. The oxygen atom is blowing in the wind. As T101 exits the oxygen atom, there is a collision with a wave of water. T101 now becomes a slow sound tadtron. He is moving in a corkscrew across the ocean. He goes for a thousand miles. The wave comes up on shore, our tadtron changes state and he is now looking for something to do.

T101 goes out past our moon and finds the end of a gravity stream. He gets to the end of the stream and follows it towards the center of the earth. The gravity goes through a space station and causes a slight push. This push helps hold the spaces station in orbit. Our tadtron continues on to the center of the earth. After it exits the atom it is going to, it fills a magnetic rib in the earth's magnetic field. The magnetic rib hits an inner magnetic ring of a magnet that is part of a compass. The contact causes the magnetic boomerang repel effect. The effect helps push the compass to point north. T101 is again looking for something to do.

He cycles through the sun as gravity and comes back to the earth as an ultraviolet light tadtron. He hits the skin on a sunbather and help cause a sunburn. T101 is now looking for something to do. He finds a gravity stream, converts to gravity, goes to a magnet and converts to an inner magnetic ring. T101 goes through a generator and is converted to an electron. The electron goes through the electricity grid. He travels to your house. He goes to a light bulb in your house. He is converted to green light. The green light hits this piece of paper. The green particle is reflected from the white part of this piece of paper and goes to your eyes. T101 is converted to an electron that goes to your brain. Your brain blends all the tadtrons and tells you that spot on the paper is white. Now you can see! T101 now changes his state and leaves your brain to continue on his never ending journey.

How many tadtrons did we start with? We had a tadtrillion tadtrons. How many did we have when we finished? We have a tadtrillion tadtrons. Energy changes states. It never goes to a state of disorder. There is always the same amount of energy, regardless of what state it is in. How many different energy states are there? I don't know. We have only identified a few. Science is always finding new kinds of energy. We may have identified less that 1 % of the energy states. Do you know what that means? There may be 99 % more energy states to identify. We may have more left to do in science.

Practical Applications

Practical Applications

Practical Applications: Opening comments

I hope you liked my ideas and theories. These ideas could improve the quality of life on our planet. What are some practical applications that can come out of this theory? One thing is better, cleaner and cheaper sources of energy. A second thing is travel beyond the speed of light. I think both of these are achievable. I am going to give some basic ideas to achieve this. Any energy source that can be converted to electricity, I am giving the suffix "verter". This will classify all the different types of converters.

The Lightverter

The Lightverter has already been invented. It has been around for many years. A lightverter converts light to electricity. The common name for a lightverter is a photocell. Photocells are used in solar panels to provide electricity to drive cars and for energy in homes. Photocells are also used in burglar alarms and to control opening and closing of elevator doors. Photocells just sit there and convert light to electricity. If this can be done with light, then other states of energy should be able to be converted as well.

The Graverter

This is a gravity converter. An example of a gravity converter is the sun. The sun converts its personal gravity to light and also produces electromagnetic energy. We should be able to do the same thing. We should be able to build a device that converts personal gravity exiting through matter and change it to electricity. It is now theoretically possible to get energy from any matter.

The Magverter

This is a magnetic converter. This has already been invented. It is currently called a generator. A generator has to move a magnet past something made of metal. A current or electricity goes into the metal. We have to move the magnet to get the current. This is where most of our electricity comes from. It may be possible to get magnets to produce an electrical current without moving the magnet. If we were able to do this we could get more energy out than what we put in. I have seen some drawings of inventions that are suppose to be able to do this. I have never seen a working model of one. It is theoretically possible to do this now.

The Freeverter

This is a free energy converter. Free energy is tadtrons in an unknown state. There may be thousands or millions of unknown states of energy. There may be thousands or millions of ways to convert this energy to electricity. Thunderstorms convert unknown energy to lightning or electricity. If a thunderstorm can do it, we should be able to do it too.

Comments On Converters

We have been raised with preconceived ideas. One of those ideas is, you can't create energy. That I agree with, however you can convert energy to a usable state. The sun does it every day. Many people are told they can't do something and they believe it. Don't believe everything you hear. If you have an idea, your 1% of the way there and 99% of the work is left to do. If you think you can build a converter, do it. It will greatly benefit mankind. It's easy to go to the gas station and fill a car up with gas. It costs money. It easy to turn a light switch on and have light, but you have to pay your electric bill. It's a bargain, but it still costs money. These energies all pollute our environment in one way or another. We can put solar panels on our homes to cut energy costs and consumptions. We can put wind mills up that generate clean electricity. There are other options to reduce the amount of energy we use. These cost money and take effort. When the right devices are built, everyone could produce their own electricity, power their own homes and power their own car.

Motion Laws

There is a set of motion laws for molecules and matter and a different set of motion laws for energy. Let's compare the 2 sets of laws.

The motion laws of molecules and matter were written by Isaac Newton.

The first law is an object stays where it is unless something acts on it. The second law is an object will move based on a force applied to the object. The third law is that when something happens to an object there is an equal and opposite reaction.

Let's use a baseball for an example as the object. If a baseball is laying on the ground it won't move unless something moves it. If someone picks up the baseball and throws it, it will move based on how hard it was thrown. The baseball will continue in the direction it was thrown until something slows it down or stops it. If the baseball hits someone, they will be knocked backward from the force of the ball. This is how the motion laws work with matter.

Visible light has a different set of laws. Light goes from one object to the next at the speed of light. If light hits matter, it has no effect on the object it hits. Light can bounce off something and go in the opposite direction at the speed of light. I named this property of light Super Reflectivity.

Gravity has a different set of laws than light. Gravity goes through objects and causes them to be pushed in that direction. It doesn't seem to bounce off or be reflected by any common objects. The only thing that seems to affect gravity is the atom it goes through. After gravity passes through an atom, it changes to something else. Each energy state may have its own set of laws. Energy seems to be able to go anywhere and at different speeds.

Let's compare visible light to a baseball. A baseball will sit there unless something moves it. Light is always in motion. A baseball has personal gravity and light doesn't. Could this be why a baseball and light are different? Let's think about that. A baseball has gravity streams coming in from all directions. Is this what holds the baseball in place? When the baseball moves, the gravity streams have to adjust. Could this adjustment be what causes momentum and inertia? It could be. Are gravity streams what causes matter to have mass?

If I move the baseball forward, it causes extra collision with the gravity streams in front of the ball. All the gravity streams from the side and back have to readjust or catch up. If the ball is slowed down or stopped, all the gravity streams from behind will push forward on the ball. All the other gravity streams will have to readjust. Is this what causes the motion laws of matter? I think it is.

What if you removed the personal gravity of the baseball? What would happen? Is it possible that matter could become like light? It would have no mass. What would that mean? It would mean that any force you applied to that matter, could cause it to accelerate to any speed? The motion laws would no longer apply to the baseball. If you were in a vacuum and you threw the ball, a small throw could cause it to accelerate faster than the speed of light. It could be many times the speed of light. What if the ball hits something? What would happen? It has no mass and no inertia. It would simply bounce off or stop. The baseball would become super reflective just like light.

Is it possible to stop the effects of gravity on matter? I think it is. The reason I believe this is based on a documentary I saw on TV. A man in England had bought electromagnetic equipment from a battleship that was being scrapped out. I don't know the individual's name, but he set the electromagnetic equipment up in his home. It was his lab. He had a special arrangement for all his equipment. When he turned it on, several objects seemed to defy the motion laws of physics. This individual filmed his experiment. On the film, objects would be sitting still, then they would wiggle and go straight up in the air. They appeared to no longer be affected by gravity.

The coolest thing he experimented with was a glass full of water. The glass wiggled a couple of times then all the water flew out the top. The water held the basic shape of the glass as it flew out. The glass wiggled a couple of more times and then it flew upwards. The individual tried to improve upon his invention, but instead of improving it, he broke it. He was then unable to reproduce his previous experiment. Several scientist viewed the film. They did not believe it was a fake. I don't think it was fraud. Ever since I viewed this film, I have believed that anti-gravity is possible. Let's see if we can do something with this.

It was some kind of electromagnetic field that produced the effect. I am going to rename the effect. I am going to name the effect a gravity cloak. A gravity cloak will block all personal gravity and the effects of gravity from all other sources. Once you have a gravity cloak around you, the motion laws of physics will no longer apply. You should be able to go any speed you want. The device that produces the gravity cloak, I am going to name a gravity cloaking engine. For simplicity, I may also refer to it as a cloak engine.

The cloak engine will produce a field of energy that nullifies the effects of gravity. There may be two ways that this can be done. One way is to produce a shell that is so thick that gravity can't go through. If we do that, it will take a large amount of energy. A second way is to fool the gravity into thinking it has gone through its atom and is now changing its state to something else. The second way is how I think it should be done. On this concept I will show plans of an anti-gravity spaceship.

Three Gravity Cloaking Engines Engaged

The illustration on the next page has the 3 parts of what will be needed to build a gravity cloak. A gravity cloak must completely enclose the object it is going to cloak, including the gravity cloaking engines.

I believe the gravity cloak will be a form of magnetism. Magnets have inner magnetic rings and magnetic ribs. A gravity cloak is going to be similar. A gravity cloak must have an inner cloaking ring and cloaking ribs. That's what I have on the next page. There are 3 cloaking engines. The object at the top is the upper cloaking engine. The object in the middle is the main cloaking engine. The object at the bottom is the lower cloaking engine.

Each of the gravity cloaking engines has a center ring. This center gravity cloaking engine ring produces the gravity cloaking ribs. What will the gravity cloaking ring be made of? I don't know. It will have to conduct some form of electricity or unknown energy. It could be a super conductor. It may conduct an energy we have not identified. The gravity cloaking engine rings will have to be inside, or very near to, the object that is going to be cloaked. Who will invent a working gravity cloak?

The gravity cloaking engines will have to be engaged. Once they engage, energy will circle inside the gravity cloaking ring. Gravity cloaking ribs will form and circle through the gravity cloaking ring. This is the same concept as a magnet. The ribs will be held in place by the rings. It may, or may not, take a massive amount of energy to fill the ring and the ribs. The illustrations show the rings and the ribs. The gravity cloaking engines are engaged. Gravity can't pass through gravity cloaking ribs. Once a gravity tadtron hits a gravity cloaking rib, it is either changed or reflected. Changing gravities state may be very easy to do. How to do it is the big question?

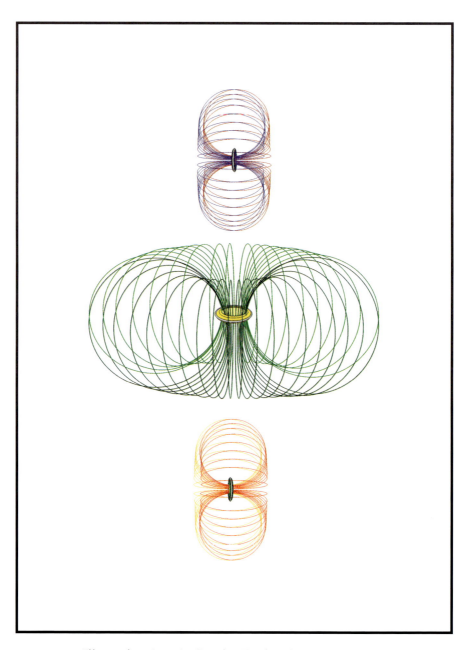

Illustration App-1: Gravity Engine Component Parts

The Galactic Cruiser

Below is an illustration of a Galactic Cruiser. The galactic cruiser will hold the gravity cloaking engine rings in place. The upper cloaking engine is located in the top center of the ship. The main cloaking engine is in the center. The lower cloaking engine is located at the bottom center of the cruiser. All the engines are held in place by crossbeams.

The engines are disengaged in this illustration. The galactic cruiser will have to provide the power to engage the cloaking engines. The light and speakers, on the hull, show the top is tilted slightly toward us. Why does the galactic cruiser have the shape it does? When we look at the next illustration, the answer will be very obvious.

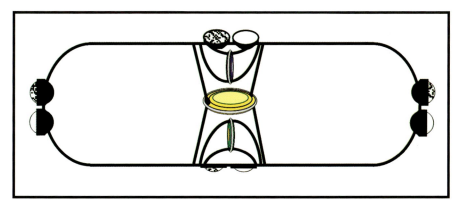

Illustration App-2: Galactic Cruiser

The galactic cruiser with gravity cloaking engines engaged

The illustration below is the galactic cruiser with the cloaking engines engaged. When the cloaking rings are turned on, the cloaking ribs form. The main cloaking engine forms a huge donut shape. The upper and lower cloaking engines fill in the holes in the donut and cover each other and the outer edge of their cloaking ribs. The shape of the galactic cruiser fits inside of the donut-shaped cloak.

The galactic cruiser is now cloaked. It doesn't have personal gravity holding it in place. In a sense, it now has no mass. It can now do any speed. The galactic cruiser is now like light, it is super reflective. How do you move the galactic cruiser? Part of the answer is in the next illustration.

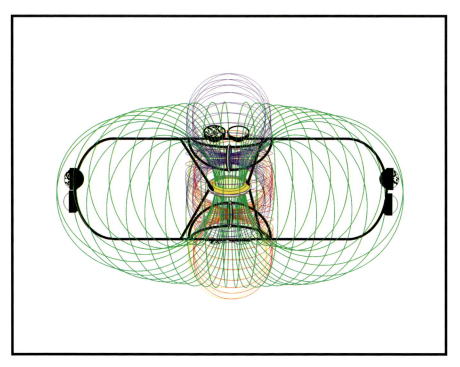

Illustration App-3: Galactic Cruiser with cloaking engines engaged

The galactic cruiser with sound and light acceleration engines

How do you make it go? There may be three ways. One way is the same way a rocket engine works. When a rocket expels its mass, it causes an equal and opposite reaction. This propels the rocket. We can't expel mass from the galactic cruiser, because it has no mass. The galactic cruiser is similar to energy. What if we expel energy like a rocket? Would exiting energy cause an equal and opposite reaction? I think it would. This is a new realm in physics. It's unchartered territory.

Could exiting sound and light propel a gravity cloaked galactic cruiser? In the illustration below the black is sound exiting the galactic cruiser. The sound will cause the galactic cruiser to push in the opposite direction from where the sound exits, producing a slight push. The yellow is light exiting the galactic cruiser. Light is very fast, it will produce a hard push on the galactic cruiser. The exiting sound or light will cause the galactic cruiser to accelerate in the opposite direction.

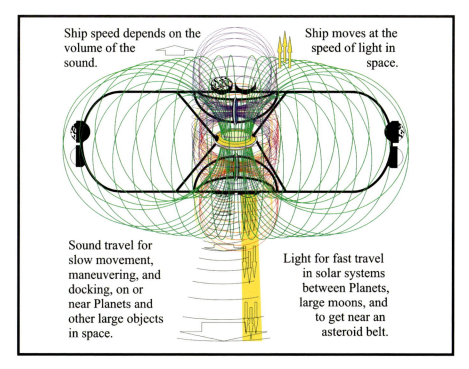

Illustration App-4: Galactic Cruiser, sound and light acceleration engines

The galactic cruiser with gravity acceleration

Is there another way to move the galactic cruiser? Yes, there is. How? Let gravity push it. How do we do that? Disengage either the upper or lower gravity cloak engine. The illustration below has the galactic cruiser with the upper and main gravity cloak engines engaged. The lower gravity cloak engine is turned off. What does this do? It lets gravity come in from the bottom. What does gravity do? It pushes everything in its path.

The arrows are gravity coming in the bottom of the ship. There is no gravity to hold the galactic cruiser in place. The push is in one direction. Gravity will cause the galactic cruiser to accelerate in that direction. Gravity may push the galactic cruiser to speeds that are unimaginable. How do we stop? Engage the lower gravity cloak engine and then disengage the upper gravity cloak engine. This should cause the galactic cruiser to slow down at the same rate it accelerated.

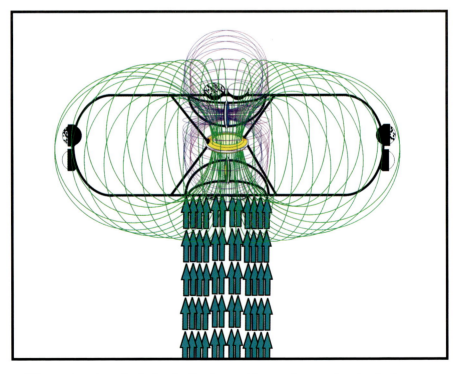

Illustration App-5: Galactic Cruiser, with gravity coming in the bottom

Practical Application: Closing Comments

How will the Galactic Cruiser act? Let's take 2 things, a bowling ball and a balloon, that are the same size. Let's put gravity cloak engines on a bowling ball. When the gravity cloak engines are disengaged, the bowling ball will act normal. When the gravity cloak engines are engaged, the bowling ball will act more like a balloon. In reality, it will have no mass and be much lighter than the balloon. If you hit the balloon, it will bounce off very rapidly. It will go a few inches and then slow down. If the balloon were in a vacuum, it would continue at the speed you hit it. If the end of the balloon was open, air would come out. The balloon would move in the opposite direction of the hole. The sound and light drives would do the same to the galactic cruiser. If you have a strong wind it will blow the balloon at that speed. Disengaging the gravity cloak on one side of the Galactic Cruiser will push it the same as the wind pushes the balloon.

The Galactic Cruiser may seem like an unbelievable idea. Let me give you a little perspective. When they were building the first airplanes, most people scoffed in disbelief. People were not bending over backwards to give them any help. When they started planning to put people on the moon, most people scoffed in disbelief. Who looks foolish now? The scoffers or the doers. Flying faster than the speed of light is doable. Converting unknown energies to electricity is doable. It takes a special person to be a doer. Anyone can be a scoffer. How fast can we go? Is it possible we could go a million times the speed of light? Is it possible we could go a billion times the speed of light? Who will travel that fast first? The doers will.

There is an unlimited number of applications possible with this theory. We could make huge gains in our knowledge of chemistry and physics. We may learn how to construct elements. What if we could tell a large group of tadtrons to form into gold? What if we could tell a massive number of tadtrons to form proton rings? Gravity would fill in the electron rings. If you could do that you could form a planet. It could be possible to form a planet in less than a second. It could be possible to form a sun in less than a second. It might take longer to form a galactic super mass or a universal super mass. Think big. Think unlimited.

The Author's Source

The Theory of Energy States is a new way of looking at energy, matter and the universe. We have a big question left. This may be the hardest question in physics. Where did the tadtrillion tadtrons come from?

Where does it all start? Is there a beginning? Is there an end? Did something always exist? If yes, was that something intelligent. Was there a big bang? These are some of the hardest questions to ask. When I was a child, I would think about there being no beginning of the universe. I would think about there being no end of the universe. This would bother me a great deal. I put this on the shelf for many years. As time passed, I had to deal with it. I did this by narrowing it down to a couple of questions. Did something always exist? There are two answers, yes or no.

If you answer no, nothing always existed. If nothing existed, nothing made everything that is currently in the universe! How can nothing make something? The big bang theory was an attempt to explain how nothing could create something. The idea behind the big bang was that nothing exploded. When nothing exploded it caused a positive universe and a negative universe of positive and negative matter. The negative matter was called anti-matter. There was also positive energy and negative energy. The negative energy was called anti-energy. When you add all the positive, to all the negative, the sum will be zero or nothing. You add all the matter and all the energy and it comes up to be equal to all the anti-matter and anti-energy. The idea is $1 + -1 = 0$. It works out mathematically. I have never seen any anti-matter or anti-energy, except on Star Trek. We all have different minds. We all think differently. My mind does not accept the big bang theory. My mind does not accept that nothing always existed. So, when we ask the question, did something always exist? I have to answer yes.

If you answer the above question yes, something always existed, was that something intelligent? You have two answers, yes or no. Let's consider answering no to this question. It would mean that what we now have in the universe has always existed, with no intelligence. So, how did it make life? By random chance? We know life on our planet has not always existed. That would mean the molecules and energy made us by random chance. Life would be a creation by random chance. Is that possible? Many believe it is. Let's look at this logically.

Is mankind, with all our combined collection of intelligence and knowledge, capable of building a life form? No, we are not. Is mankind capable of building tadtrons? No, we are not. Let's work our way down. Let's go with something with less intelligence and with less knowledge than all mankind. Let's go with YOU. Are you capable of building a life form? No, you are not. Are you capable of building tadtrons? No, you are not. Let's continue to work our way down. A house is less complicated than life or tadtrons. Are you capable of building a house? You probably are. Let's continue to work our way down further. A dog has less intelligence and less knowledge than you. Can a dog build a house? No, it can't. Let's continue to work our way down even further. "Nothing" has less intelligence and less knowledge than a dog. Can nothing build a house? No, it cannot. Can nothing build life? No, it cannot. Can nothing build the tadtrons? No, it cannot. What's my conclusion? My mind tells me that something less intelligent than my dog could not have made the universe what it is. There is something around that is more intelligent than mankind.

If you still believe in a big bang, here's the math: 1 tadtrillion tadtrons plus 1 tadtrillion anti-tadtrons equals 0. Everything equals nothing. Everything came from nothing. Everything you see, hear, smell, taste and touch is really nothing.

This is where I stand. I believe that "something intelligent" has always existed. I believe that "something intelligent" made everything in this universe. I believe that "something intelligent" made life. I believe that "something intelligent" made the tadtrons. What are the tadtrons? Tadtrons are a brilliant design. They are single particles that can go anywhere and can become anything. What are tadtrons? Tadtrons are the signature of a BRILLIANT CREATOR!

Dannel Roberts is also the successful author of an ongoing adventure series of "Me and Uncle Mike" children's books.

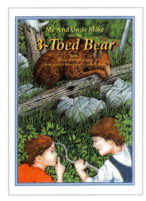
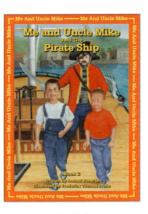
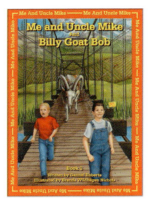

Book 1
Me and Uncle Mike
and the
3-Toed Bear

Book 2
Me and Uncle Mike
and the
Pirate Ship

Book 3
Me and Uncle Mike
and
Billy Goat Bob

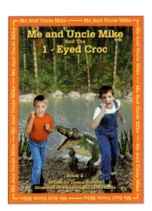

Book 4
Me and Uncle Mike
and the
1-Eyed Croc

Book 5
Me and Uncle Mike
and the
Purple Gorilla

These books are 9 x 12 , large print editions, easy to read for ages 2 through 8. Every other page displays full-color, professional quality artwork that enhances the fun and adventures of two young boys.

Text, color image examples and online ordering available at
www.meandunclemike.com